Water management in Islam

The UNU Programme on Integrated Basin Management focuses on water management, approaching the complex problematique from three particular angles: governance, capacity-building, and management tools. The programme is carried out through field-based research encompassing both natural and social sciences. It utilizes extensive networks of scholars and institutions in both developing and industrialized countries. This work is intended to contribute to policy-making by the United Nations and the international community, as well as to capacity-building in developing countries.

The Water Resources Management and Policy series disseminates the results of research carried out under the Programme on Integrated Basin Management and related activities. The series focuses on policy-relevant topics of wide interest to scholars, practitioners, and policy-makers.

Earlier books in this series are:

Hydropolitics Along the Jordan River: Scarce Water and Its Impact on the Arab-Israeli Conflict by Aaron T. Wolf

Managing Water for Peace in the Middle East: Alternative Strategies by Masahiro Murakami

Freshwater Resources in Arid Lands edited by Juha I. Uitto and Jutta Schneider

Central Eurasian Water Crisis: Caspian, Aral, and Dead Seas edited by Iwao Kobori and Michael H. Glantz

Latin American River Basins: Amazon, Plata, and São Francisco edited by Asit K. Biswas, Newton V. Cordiero, Benedito P. F. Braga, and Cecilia Tortajada

Water for Urban Areas: Challenges and Perspectives by Juha I. Uitto and Asit K. Biswas

Water management in Islam

Edited by Naser I. Faruqui, Asit K. Biswas,
and Murad J. Bino

TOKYO • NEW YORK • PARIS

International Development Research Centre

Ottawa – Cairo – Dakar – Johannesburg – Montevideo – Nairobi – New Delhi – Singapore

Published in Europe and the United States of America by the United Nations University Press
The United Nations University, 53-70, Jingumae 5-chome,
Shibuya-ku, Tokyo, 150-8925, Japan
Tel: +81-3-3499-2811 Fax: +81-3406-7345
Email: sales@hq.unu.edu
http://www.unu.edu

Published in Canada by the International Development Research Centre
PO Box 8500, Ottawa, ON, Canada K1G 3H9
Tel: +1-613-236-6163 Fax: +1-613-563-2476
Email: pub@idrc.ca
http://www.idrc.ca/booktique/

Library of Congress Cataloging-in-Publication Data

Library of Congress Cataloging-in-Publication Data

Water management in Islam / edited by Naser I. Faruqui, Asit K. Biswas, and Murad J. Bino.
p. cm.
Based on the findings of the Workshop on Water Resources Management in the Islamic World, held in Amman, Jordan in Dec. 1998, organized by the International Development Research Centre. Includes bibliographical references.
ISBN
1. Water-supply-Middle East-Management. 2. Water-supply-Africa, North-Management. 3. Water-supply-Islamic countries-Management. 4. Water resources development-Government policy-Islamic countries.
I. Faruqui, Naser I. II. Biswas, Asit K. III. Bino, Murad J. IV. International Development Research Centre (Canada) V. Title.
TD313.5 .W38 2000
363.6′1′0917671—dc21 00-011417

Canadian Cataloging in Publication Data

Main entry under title:
Water management in Islam

"This book is primarily based on the findings of the Workshop on Water Resources Management in the Islamic World, held in Amman, Jordan, in December 1998". — Pref.
Co-published by UNU Press.
ISBN 0-88936-924-0

1. Water resources development — Middle East.
2. Water supply — Middle East.
3. Water resources development — Africa, North.
4. Water quality management — Middle East.
5. Water quality management — Africa, North.
I. Faruqui, Naser I., 1965– .
II. Biswas, Asit K.
III. Bino, Murad J.
IV. International Development Research Centre (Canada)
HD1698.7W27 2000 333.91′00956 C00-980407-2

Cover design by Joyce C. Weston.
Cover photograph by Murad Bino.

ISBN 92-808-1036-7 (UNUP edition)
ISBN 0- 88936-924-0 (IDRC edition)

Contents

Foreword

The theory and practice of natural resources management is changing. The formerly neat and isolated compartments in the study of natural resources management are disappearing. In universities, scientists are starting to analyse natural resources from multidisciplinary perspectives. In the field, resource managers are being urged to take account of the concerns of the full range of stake-holders. In both public and private sectors, decision makers are recognizing forces which, until recently, they considered as extraneous. One of these is globalization which brings foreign institutions and actors into the domestic arena. Another is the re-introduction of explicit value systems into policy choices.

Many value systems have their origins in religion. Recognizing the role of religion can therefore enrich our understanding of how individual and collective choices are made, independently of our own belief in, or attitude towards, religion in general or towards a particular religion. *Water Management in Islam* presents interpretations by eighteen scientists of the role that Islam may play in water management. Each of the authors has great experience in one or another aspect of the topic. Since Islam is the religion of about one-fifth of the world's population and the official faith of a number of countries, in many of which water is the key scarce factor for development, understanding its actual or potential role is important.

This book makes a valuable contribution to development by presenting the Islamic perspective on a number of proposed water management policies such as water demand management, wastewater reuse, and fair pricing. These policies are widely agreed to lead to more equitable, efficient, and sustainable water management. While water management practices and policies are influenced by a combination of social, cultural, economic, and political factors, not all of which relate to religious value systems, the congruence between Islamic principles and those currently promoted, such as the Dublin Principles, is worth noting. It is much closer than many theorists and practitioners realize. The book thus opens avenues for a wider dialogue among researchers working at identifying the most promising water management policies and for a more rapid adoption of these. It also adds to our knowledge of some of the influences on formal policy and informal practice and makes these ideas available to a broader public.

Caroline Pestieau
Vice President, Programs
International Development Research Centre
Ottawa, Canada

Preface

This book explores the Islamic perspective on a number of proposed water management policies, such as lifeline water tariffs, water conservation, wastewater reuse, community-based water management, fair pricing, and water markets. These measures are generally accepted, with certain provisos, to lead to more equitable, efficient, and sustainable water management. By studying these issues in the context of Islam, workshop participants were able to derive Islamic water management principles that were in harmony with currently accepted principles of sustainable water management.

The book is primarily based on the findings of the Workshop on Water Resources Management in the Islamic World, held in Amman, Jordan, in December 1998. The workshop was organized by the International Development Research Centre (IDRC), with support from the International Water Resources Association (IWRA), and the Inter-Islamic Network on Water Resources Demand and Management (INWRDAM). However, the discussion and conclusions presented reflect the consensus and interpretations of the participants in the workshop. They do not reflect official policies of IDRC, IWRA, or INWRDAM.

The mission of the IDRC, a Canadian Crown Corporation, is "empowerment through knowledge," and its mandate is to initiate, support, and conduct research to help communities in the developing world find solutions to their social, economic, and environmental problems. One IDRC program, entitled "People, Land, and Water," focuses on research that helps people in the MENA region to better manage their land and water resources. A specific objective of this program is to contribute to local and national policies and institutional arrangements that equitably increase the quality and accessibility of water resources. Furthermore, IDRC follows a people-centred approach that requires staff and research partners to examine each research problem from the perspective of the beneficiaries' socio-economic circumstances.

Because most of the countries within the MENA are predominately Muslim, and because it has been suggested that Islam is against some currently promoted water management policies, IDRC organized the workshop with the following goal:

> To develop a better understanding of Islamic perspectives relating to selected water management practices and outline research necessary to develop water management policies that will improve the lives of the poor.

This book is expected to be useful to researchers, policy-makers, donor organizations, and non-governmental organizations working in countries with largely Muslim populations. Although the workshop participants were able to agree upon water management principles that could be considered Islamic, these principles are not unique. As one delves into Islam, one encounters values common to the other two Abrahamic monotheistic religions, Christianity and Judaism, whose holy books Muslims recognize. Thus, a water conservation project incorporating local values in, for example, Egypt, which has a large Christian minority, could include verses from the Bible, as well as those from the Quran and *hadiths*, which would complement each other. Because these values are universal, they are not common only to these closely related faiths, but are also inherent in many other belief systems.

Beyond the specific value of this book for water management in the Middle East and North Africa, it serves as a concrete example of the benefit of examining development in the context of values and culture. This approach is consistent with the International Development Research Centre's philosophy of examining development research problems from the perspective of its southern partners and engaging them in the research process. However, examining values is not easy. Most scientists and even development professionals avoid examining religion or values in the context of their work to avoid discord and to keep discussion "objective." However, where science, development, and values intersect, the issue cannot be avoided. For example, in the past, it may not have been scientifically possible to treat wastewater to the extent that it could be safely reused but, under certain conditions, this can now be done. What does a religion such as Islam, which, like other belief systems, puts such a premium on purity, have to say about this?

Examining values can be particularly sensitive in the Middle East, where the prevailing religion is Islam, but which also has many adherents of other faiths, such as Christianity, Zoroastrianism, and Judaism. However, in a spirit of tolerance and mutual respect, workshop participants reached consensus that followers of different belief systems have much to learn from each other.

Although the concrete outputs of this work are relevant to water management in the Middle East and North Africa, more broadly the exercise demonstrates the value of developing a deeper understanding of cultures and belief systems other than one's own. It may be challenging, and even sensitive, but it is worth it. Take for instance, the participation of the Roman Catholic Church in recent discussions on human rights in Latin America. For many Catholics in the region, the Church legitimized the debate, by adding to its moral dimension, and by emphasizing the role of the family and human responsibilities as well as human rights. We hope that this exploration of Islam and water will lead to the examination of other belief systems in different development contexts.

Naser I. Faruqui
Asit K. Biswas
Murad J. Bino

Acknowledgements

To edit any book on a sensitive subject such as religion is always challenging. To make concrete statements on a subject – Islam and water – that has been little explored in the past is doubly challenging. In addition to God, I would like to thank the following individuals, without whose help it would not have been possible to publish this book: my assistant Saeeda Khan, for her enthusiastic help organizing the workshop and editing the book; Abdoul Abdoulaye Sow, for continuing where Saeeda left off when she returned to school; Francis Thompson, Dr. Gamal Soleiman (the Imam of the Ottawa Mosque), and my colleagues at IDRC for their valuable comments in reviewing the manuscript; Yassine Djebbar, Ghazi Al Nakshabandi, Odeh Al-Jayyousi, Hussein Amery, and Alauddin Ahmad for the painstaking task of reviewing the glossary, providing English translations of the titles of Arabic works, and referencing *hadiths*; and my co-editors, and the organizations that they represent, Dr. Asit Biswas (IWRA) and Dr. Murad Bino (INWRDAM). All of the participants in the workshop in Jordan and particularly the authors contributing to this book deserve gratitude for their submissions and for their input into and review of the overview chapter. Finally, I thank my wife Natasha for her unwavering support while I worked on this task.

Naser I. Faruqui

Introduction

Naser I. Faruqui

In the Middle East and North Africa (MENA),[1] water is rapidly becoming the key development issue. The region has one of the highest average population growth rates in the world (around 2.8 per cent), and scarce natural water supplies. As a result, renewable available water in the region dropped from an average of 3,300 cubic metres per person per year (m^3/p/y) in 1960 to 1,250 m^3/p/y in 1996, and is expected to decline to 725 m^3/p/y by 2025.

Many countries in the region already fall well below a water availability of 500 m^3/p/y. For example, in the early 1990s, the annual renewable freshwater available per person in Jordan, Tunisia, and Yemen was 327, 540, and 445 m^3, respectively, and these values are projected to drop drastically by 2025 (World Bank 1995). Almost all of the states of the Arabian peninsula, in addition to Israel, Jordan, and Libya, already consume much more water than their annual renewable supplies. Egypt, Morocco, Sudan, Syria, and Tunisia are fast approaching the same critical threshold. Furthermore, the available water is of lower quality because of increasing pollution and overpumping. A benchmark level of 1,000 m^3/p/y is often used as an indicator of water scarcity: below this, a country is likely to experience chronic water scarcity on a scale sufficient to impede development and harm human health (Falkenmark and Lindh 1974) – 500 m^3/p/y indicates severe water stress.

Competing water demands are exacerbated by high population growth rates and rapid urbanization. Although the collective urban growth rate of less developed countries (LDCs) is estimated to be 2.9 per cent for the period 1995–2015, it is even higher in MENA countries – 3.2 per cent. By 2015, the proportion of the total population living in urban areas will be 66 per cent, compared with 49 per cent for LDCs as a whole (UNDP 1998).

Against this backdrop, national governments, non-governmental organizations (NGOs), and donor organizations, including IDRC, are working to meet the challenge of providing all of the people in the MENA region with an adequate supply of freshwater to meet their needs.

Values and development

Culture, including religion, clearly influences how people perceive and manage a resource such as water. Although this aspect was often neglected in development projects in the past, development agencies increasingly acknowledge the importance of local culture and values in their policies. However, this is not necessarily reflected in their projects. Projects that incorporate culture either tend to focus on certain small groups, primarily indigenous groups in rural areas – for example, the relationship between the belief system of Masai in Kenya and their nomadic herding lifestyle – or on specific elements, such as the effect of male-dominated societies on female literacy. Although these studies are useful, donor organizations tend to focus on particular issues, such as gender equity, without examining the broader cultural context. Remarkably few studies of this nature have been carried out in the Middle East, nor, for example, on the impact of Catholicism on family planning policies in Latin America, or of Hinduism on soil management in India. There seems to be a belief that comprehensive in-depth studies that look at all aspects of a particular religion or culture are becoming unimportant in a globalizing, urbanizing world, in which religion or beliefs are perceived to be less important as the world proceeds toward some kind of common, material-based culture.

Yet, researchers, policy-makers, and donor organizations are beginning to acknowledge the failure of development projects that ignore local values. The United Nations Environment Programme (UNEP) commissioned a collection of essays on ethics because "if the Rio consensus is to have any meaning, it has to be grounded in the notion of human right and wrong in relation to the way the earth and its bounty are made to serve the interests of all" (UNEP 1994, 2). The Food and Agriculture Organization (FAO) of the United Nations first looked at the issue of water rights in Muslim countries in 1954, and later published a study on *Water Laws in Moslem Countries* in 1973. In 1996, the World Health Organiza-

tion (WHO) published a series of booklets on health education through religion, addressing subjects such as water, sanitation, and environmental health in Islam. In 1995, the World Bank helped sponsor a conference on "Ethics and Spiritual Values: Promoting Environmentally Sustainable Development." It is currently consulting with leaders from nine of the world's faiths to broaden understanding and action in tackling the issue of global poverty. The Canadian International Development Agency (CIDA) hosted a dialogue on "Spirituality in Sustainable Development" in 1996. The recently established World Water Commission (WWC), which includes senior representatives of the World Bank, Rockefeller Foundation, Carnegie Endowment, World Conservation Union, and the Arab Fund for Economic and Social Development, is investigating the relationship between religion and water management practices.

IDRC has an ongoing project on science, religion, and development, which follows from a study on culture, spirituality, and economic development. The study is intended to help IDRC better integrate human values and belief systems, within its mandate, in present and future development work (Ryan 1995). IDRC has in the past also sponsored or co-sponsored several studies on values, including a 1988 Conference on Ethics and Human Values in Family Planning in Thailand, a 1985 study of the effect of traditional and religious beliefs on the acquisition of knowledge in Ethiopia, and a 1976 thesis on religion, kinship, and labour and their effects on Luapulan fishing communities in Africa.

Islam in the MENA

Examining underlying values is particularly important in the Middle East and North Africa, which, along with large minorities following various faiths, is home to three hundred million Muslims. As with some other belief systems, Islam encompasses much more than worship and a code of personal conduct implied by the word "religion" (Bankowski et al. 1988, 222). It regulates virtually all aspects of individual and collective human life, for example, issues such as buying and selling, contracts, inheritance, marriage, family and intimate relations, and even elemental issues such as eating and personal hygiene and sanitation.

Within Islam as a religion,[2] the concept of "secularism," or separation of mosque and state, does not exist. Although only a handful of Muslim countries now base their political, judicial, economic, or constitutional systems completely on Islamic law, the influence of Islam is strong enough to preclude calling even "Westernized" countries, such as Tunisia and Morocco, secular. In fact, with the exception of Turkey, it is difficult to identify a purely secular state in the Middle East. In contrast, appeals

to religious values, predominately Islamic, are common, even in relatively secular countries, in a way that today would seem remarkable in a Western country. For instance, a municipal water utility in Canada, such as the Greater Vancouver Regional District, would never quote from the Bible to promote water conservation. It is unlikely that this would have happened even in the far less secular climate found in Canada as recently as thirty years ago. Yet in Jordan, a country which is relatively secular compared to Iran or Saudi Arabia, and which has significant religious minorities, the Ministry of Water uses Islamic sources along with secular slogans to promote water conservation by the population.

Furthermore, since about twenty years ago the influence of Islam has grown in the region. The severe challenge posed by the MENA's dwindling per capita water availability suggests two things:

- Policy-makers will have to use all tools available to address it, including those neglected in the past, such as culture and values; and
- Increasing competition among sectors and overlapping claims to the same water among individuals and groups mean that decisions and sacrifices will have to be made at the local level, and the nature and extent of these sacrifices will depend upon personal and communal values.

The Dublin Principles

The emerging international consensus on water management was most recently outlined at the 1992 UNEP Water Conference held in Dublin, Ireland.[3] The often quoted principles agreed upon at the conference are as follows.

- Water is a social good;
- Water is an economic good;
- Water management ought to be participatory and integrated; and
- Women play a central role in water management.

Because these principles or objectives are so general, water management professionals have identified practices or policies that can help achieve them, such as lifeline water tariffs, water conservation, raising tariffs, wastewater reuse, privatization, water markets, and community-based water management. Policy-makers generally accept that these practices, with certain provisos, are valuable and will help promote equity. However, some Muslims have said, before and after the Dublin Conference, that Islam is against selling water and wastewater reuse. These statements have been circulated in the literature, for example "There are religions (for example Islam) that prohibit water allocation by market forces"

(Webb and Iskandarani 1998, 34). Thus, the purpose of our workshop was to examine the Islamic perspective on these proposed practices. Does Islam agree with them or not? If so, with what limitations? Without examining the underlying values in Islam, it will be difficult to achieve the objectives of the Dublin Conference, and thereby increase equity, efficiency, and sustainability of the water supply in many parts of the MENA.

The scope of the workshop

With these considerations in mind, the workshop call for papers invited detailed abstracts on four main topics and several sub-topics:

- Water as a social good –
  - importance of equity in Islam;
  - categories and priorities of rights; and
  - rights of the environment
- Non-economic instruments of water demand management –
  - water conservation in Islam;
  - public awareness; and
  - wastewater reuse and management
- Economic instruments of water demand management –
  - permissibility of trading and cost recovery in Islam;
  - water pricing;
  - intrasectoral water trade;
  - privatization of service delivery; and
  - intersectoral water markets and reallocation
- Integrated water management –
  - community participation in water management;
  - international water management; and
  - national-level policy setting

To ensure a small meeting, with high-quality papers prepared by experts in water management, who were well versed in Islam and proficient in English, only a small number of respondents were invited to develop their abstracts into full-length papers and to participate in the workshop. As a result, although the workshop did produce some new insights in the topics on which it focused, other topics, such as Islam and community-based water resources management, require further investigation.

Although the call for papers was sent all over the world, most of the respondents were from the Middle East. For this reason, and because IDRC's main water initiative, "People, Land, and Water," focuses on Africa and the Middle East, most of the examples cited are from that region. Also, given that of all the regions of the world with a Muslim majority of the population, this is the one facing the greatest water challenge, it seemed appropriate to focus on the area. At times, the terms "MENA" and "Muslim countries" are used interchangeably in this book. However, the findings are not only relevant to predominately Muslim countries, but other countries as well, and certainly to Muslim countries not found in the Middle East.

The opening chapter of this book gives an overview of the papers presented at the workshop, as well as of the consensus on principles and suggestions for further research that emerged from the discussions; the papers themselves then follow. For the convenience of readers who are unfamiliar with Islam, a brief description of Islamic sources and a glossary of Arabic and Islamic terms are also provided.

Notes

1. In this book, the term MENA is used to refer to those countries of the region in which IDRC supports projects: Algeria, Egypt, Jordan, Lebanon, Morocco, Palestine, Sudan, Syria, Tunisia, and Yemen.
2. Throughout this work, a distinction is made between Islam – the religion – and Muslims – those who follow Islam. The terms "Islamic" and "Muslim" are not synonymous because the actions of individual Muslims are not always consistent with Islamic teachings. This statement should not be interpreted as a value judgement. It is simply an observation that some Muslims are more observant and others are less observant.
3. The conference "Water and the Environment: Development Issues for the 21st Century" called for new approaches to the assessment, development, and management of freshwater resources. This need for reforms in water resources management was later confirmed at the UN Earth Summit in Rio de Janeiro, Brazil, in June 1992.

REFERENCES

Bankowski, Z., Barzelatto, J., and Capron, A. M. (eds.) (1988), *Ethics and Human Values in Family Planning: Conference Highlights, Papers and Discussion: 22nd Council for International Organizations of Medical Sciences (CIOMS) Conference, Bangkok, Thailand, 19–24 June*, WHO, Special Programme of Research, Development and Research Training in Human Reproduction, Geneva.

Falkenmark, M., and Lindh, G. (1974), "How Can We Cope with Water Resources Situation by the Year 2050?" *Ambio* 3 (3–4), pp. 114–22.

Ryan, William F. (1995), *Culture, Spirituality and Economic Development: Opening a Dialogue*, International Development Research Centre, Ottawa.

UNDP (United Nations Development Programme) (1998), *Human Development Report 1998*, Oxford University Press, New York.

UNEP (United Nations Environment Programme) (1994), *Ethics and Agenda 21: Moral Implications of a Global Consensus*, UNEP, Nairobi.

Webb, Patrick and Maria Iskandarani (1998), *Water Insecurity and the Poor: Issues and Research Needs*, Center for Development Research, Universität Bonn, Bonn.

World Bank (1995), "Earth Faces Water Crisis" (press release), World Bank, Washington, D.C.

Yusuf Ali, A. (1977), *The Holy Qur'an: text, translation, and commentary*. American Trust Publications for The Muslim Student Association of the United States and Canada, Plainfield, Ind.

Islamic sources

Naser I. Faruqui and Odeh Al-Jayyousi

Readers of this book are likely to possess some general knowledge of Islam as a religion, or to be able to find information about it elsewhere; but they are less likely to be familiar with the Islamic sources upon which our discussions, arguments, and conclusions are based. Accordingly, these sources are briefly described here.

Sharia or Islamic law, which is frequently referred to in this book, covers all aspects of the Islamic faith, including beliefs and practices. *Sharia* is derived primarily from the Quran and the *sunnah* (the way of life of the Prophet Muhammad (pbuh),[1] as known from *hadith*, or documented narrations of what the Prophet said and did), and also through *ijtihad* (inquiry, interpretation, and innovation by Muslim scholars). To be academically valid, the presentation of the Islamic perspective on any issue must be based upon these sources, particularly the first two. Therefore, to be consistent with all legitimate discussions of Islam, this book quotes both the Quran and the *sunnah* extensively.

The Quran is the primary source for Islamic values.[2] Muslims believe that it is the exact word of Allah revealed to the Prophet Muhammad (pbuh) through the Angel Gabriel. While the Quran does contain some specific prescriptions that rank as legal, primarily it establishes a general set of moral guidelines – a compass for Muslims to use in following an Islamic way of life. The Quran says of itself that "*Here is a plain statement to men,*[3] *a guidance and instruction to those who fear Allah.*"[4]

The *sunnah* reflects what the Prophet said, did, or tacitly approved. In Islam, following the example of the Prophet is given such importance because the Quran instructs Muslims to follow him: for example, "*O ye who believe! obey Allah and obey the Apostle.*"[5] For Muslims, the Prophet was the perfect human role model and leader, and the society he built around him is the model for a caring, equitable society. The degree of importance attached to the Prophet in Islam may not find parallel in some other religions. If the Quran is a compass for Muslims, the *sunnah* is a more detailed map for the human journey on this earth.

Some of the *sahaba*, or companions of the Prophet (pbuh), memorized and wrote down what the Prophet said or did. These documented narrations are called *hadith*, and they were later checked for authenticity, based upon such factors as independent confirmation, soundness of the chain of narration, credibility of narrators in the chain, and consistency with other *hadith* and the Quran. In specific cases, known as *hadith qudsi*, Muslims believe that the Prophet's sayings are the revelation of Allah expressed in the Prophet's own words. The science of *hadith* criticism (*mustalan al-hadith*) practiced by Islamic scholars also includes the examination of context: that is, is the *hadith* relevant only to the specific time and situation, or is it relevant to other cases? Of the many collections of *hadith*, six are considered to be the most accurate and reliable: those of Imam Al-Bukhari, Imam Muslim, Imam Al-Tirmidhi, Imam Malik, Imam Abu-Dawood, and Imam Ibn Majah. We have relied in this publication on the collection of *hadith* in the Alim CD-ROM for consistency and ease of follow-up research; occasionally we cite from the *Hadith Encyclopedia* CD version.[6]

In addition to the Quran and the *sunnah*, *ijtihad* can be used to make rulings that address new questions related to changing conditions. Essentially, *ijtihad* is the development of *Sharia* from its sources (that is, the Quran and the *sunnah*). The five tools used to carry out *ijtihad* are *qiyas* (analogy), *ijma* (unanimous agreement of jurists), *istihsan* (juristic preference), *maslaha* or *istislah* (public interest or human welfare), and *istishab* (continuity or permanence). Although some inquiry is indeed prohibited in Islam (*bid'ah sayyi'ah*), such as questioning key articles of Islamic faith – for example, the oneness of God – there is an increasing need for multidisciplinary creative inquiry into new problems and questions arising in today's dynamic society, following the guidelines already established by the Quran and *sunnah*. It is in the spirit of this type of inquiry that this book was undertaken.

Notes

1. "Peace be upon him" (pbuh) is an expression of respect which Muslims use when mentioning all prophets in whom they believe, including Jesus and Moses. Most of the workshop participants used this expression in their chapters, and it is used in the introductory parts of this book, including the summary chapter.

2. Throughout this book, citations from the Quran are in italics. For consistency, all Quranic references have been taken from Abdullah Yusuf Ali's translation (Yusuf Ali 1977).
3. The only authentic form of the Quran is its original Arabic version, and in the Arabic language, every word has a masculine and feminine form, with the masculine form taking grammatical precedence. Therefore, whenever a Quranic or *hadith* reference uses the term "he" in a general sense, the implication is "he or she." This is supported by the following *hadith* where a woman asked the Prophet if a general ruling he had addressed to men was also applicable to women, and he responded, "Yes. Women are the counterpart of men" (Abu-Dawood: 236).
4. 3:138.
5. 4:59.
6. The Alim CD-ROM includes the *hadith qudsi*, along with *hadith* compilations of Sahih Al-Bukhari, Al-Muwatta, and abridged versions of Sahih Muslim, Al-Tirmidhi, and the *Sunan* of Abu-Dawood (Shah 1986–96). The *Hadith Encyclopedia* also contains valuable material (Sakhr 1992).

REFERENCES

Sakhr (1992), *Hadith Encyclopedia* (CD version 2.0), Sakhr Software Company, Sakhr Building, Nasr Company, Free Zone, Cairo.

Shah, Shahid N. (1986–96), *The Alim for Windows* (release 4.5), ISL Software Corporation.

Yusuf Ali, A. (1977), *The Holy Qur'an: Text, Translation, and Commentary*, American Trust Publications for The Muslim Student Association of the United States and Canada, Plainfield, Ind.

Abbreviations

BWR	Basic water requirement
CEHA	WHO Regional Centre for Environmental Health Activities
CIDA	Canadian International Development Agency
CLIS	Council of Leading Islamic Scholars
CWRA	Canadian Water Resources Association
EMR	WHO Eastern Mediterranean Region
EMRO	WHO Eastern Mediterranean Regional Office
ESCWA	UN Economic and Social Commission for Western Asia
FAO	Food and Agriculture Organization of the United Nations
GCC	Gulf Cooperation Council
IDB	Islamic Development Bank
IDRC	International Development Research Centre
ILA	International Law Association
ILC	International Law Commission
INWRDAM	Inter-Islamic Network on Water Resources Demand and Management
IWM	Islamic water management
IWMP	Islamic water management principles
IWRA	International Water Resources Association
IWRM	Integrated water resources management
LDC	Less-developed country
LDMC	Less-developed Muslim country

LPCD	Litres per capita per day
m^3/p/y	Cubic metres per person per year
MAW	Ministry of Agriculture and Water, Saudi Arabia
MCM	Million cubic metres
MCM/y	Million cubic metres per year
MENA	Middle East and North Africa
NCWCP	National Community Water Conservation Programme (Egypt)
NGO	Non-governmental organization
NIS	New Israeli shekel (US$1 = 4.1 NIS)
pbuh	Peace be upon him
SIWI	Stockholm International Water Institute
SR	Saudi riyal (US$1 = 3.751 SR)
SWCC	Saline Water Conversion Corporation, Saudi Arabia
UN	United Nations
UNDP	United Nations Development Programme
UNEP	United Nations Environment Programme
WDM	Water demand management
WHO	World Health Organization
WRM	Water resources management
WUA	Water users association
WWA	Water and Wastewater Authority, Saudi Arabia
WWC	World Water Commission

Water management in Islam

1

Islam and water management: Overview and principles

Naser I. Faruqui

This chapter is based primarily on the papers and discussion from the Workshop on Water Resources Management in the Islamic World, complemented by further analysis and review of other sources. Following the overview, a set of Islamic water management principles is presented, along with recommendations, significant further research, and conclusions.

Water as a social good

Water is of profound importance in Islam. It is considered a blessing from God that gives and sustains life, and purifies humankind and the earth.

The Arabic word for water, *ma'*, occurs sixty-three times in the Quran. God's throne is described as resting on water, and Paradise is described as "*Gardens beneath which rivers flow*."[1] As Caponera (this volume) points out, it seems that in the Quran, the most precious creation after humankind is water. The life-giving quality of water is reflected in the verse, "*And Allah has sent down the water from the sky and therewith gives life to the earth after its death*."[2] Not only does water give life, but every life is itself made of water: "*We made from water every living thing*."[3]

All human beings rely on water for life and good health but, for Muslims, it enjoys special importance for its use in *wudu* (ablution, that is, washing before prayer) and *ghusl* (bathing). The benefit of the daily prayers, one of the Five Pillars of Islam, has itself been compared by the Prophet (pbuh) to the cleansing action of water in the following *hadith*, "The similitude of five prayers is like an overflowing river passing by the gate of one of you in which he washes five times daily."[4]

Water and equity

Muslims believe that ensuring social justice, or equity, in society is the cornerstone of Islam, and that the Prophet Muhammad (pbuh) set the example for them in this regard. Virtually all of the *hadith* relate to the preservation of equity, and those related to water are no exception. For example, "None of you will have faith till he wishes for his (Muslim) brother what he likes for himself."[5] Obviously, this applies to the desire for an adequate amount of clean, fresh water, as well as anything else. A Muslim cannot hoard excess water – rather he is obliged to allow others to benefit by it. The Prophet (pbuh) stated that among the three people Allah will ignore on the day of resurrection are "a man [who] possessed superfluous water on a way and he withheld it from the travellers."[6] The Quran warns human beings against unfair distribution by stating that the riches of this world belong to Allah, his Prophet, orphans, the needy, and the wayfarer, and that these riches ought "*not (merely) make a circuit between the wealthy among you.*"[7] In fact, the recognition of water as a vital resource, of which everyone has the right to a fair share, is emphasized by the following *hadith*, which effectively makes water a community resource to which all, rich or poor, have a right: "Muslims have common share in three things: grass (pasture), water and fire (fuel)."[8] On the Prophet's advice, one of his companions, Othman, who later became the third Muslim caliph, bought the well of Ruma (a settlement in Arabia) and made its water available free to the Muslim community – the well was actually made into a *waqf*, a usufruct or a collective property for religious purposes and public utility.

Rights of the environment

As in Christianity and Judaism, in Islam humankind has the first right to the resources that God has provided for his creation. It is well accepted by Islamic scholars (Mallat 1995, 129) that the priority of water use rights is: first, *haq al shafa* or *shirb*, the law of thirst or the right of humans to drink or quench their thirst; second, *haq al shafa*, the right of cattle and household animals; and third, the right of irrigation. However, as discussed later, the environment has clear and unmistakable rights in Islam.

God informs human beings of the rights of animals by comparing them (animals) to humans: "*There is not an animal (that lives) on the earth, nor a being that flies on its wings, but (forms part of) communities like you.*"[9] Animals cannot be allowed to die of thirst, and the water that remains after humans have quenched their thirst must be given to them. The Prophet (pbuh) said, "there is a reward for serving any animate (living) being,"[10] and "He who digs a well in the desert ... cannot prevent the animals from slaking their thirst at this well."[11] The immense value of giving water to any creature is reflected by the following *hadith*: "A prostitute was forgiven by Allah, because, passing by a panting dog near a well and seeing that the dog was about to die of thirst, she took off her shoe, and tying it with her head-cover she drew out some water for it. So, Allah forgave her because of that."[12]

The Quran notes that the gift of water is for flora as well: "*vegetation of all kinds*"[13] and "*various colours*"[14] are nourished by rainwater that God sends down.

These verses support the statement that water is made available by God so that all life should receive support according to its needs, including humans, animals, and plants (Yusuf Ali 1977, n. 3107). As Amery (this volume) notes, nonhuman species have rights to sufficient water that is of "good" quality because the water has to be suitable for "nourishing vegetation" and for drinking by animals.

Humankind's role as steward

Although humans are the most favoured of God's creation, we also are responsible for ensuring that God's gifts are available to all living things. As Amery (this volume) points out, in Islam, human-environment interactions are guided by the notion of humans as *khulafa*, viceregents or stewards, of the earth. Khalid (1996) states that although "we (humans) are equal partners with everything else in the natural world, we have added responsibilities. We are decidedly not its lords and masters: but its friends and guardians." Given that the Arabic root of Islam, *salam*, means peace and harmony, and the specific rights of the environment outlined in the Quran and *hadith*, Ansari (1994, 394) argues that an "Islamic way of life entails living in peace and harmony" at ecological, as well as individual and social levels.

The environment is protected from humans by specific injunctions against upsetting its natural order through pollution or other activities. In the Quran, Allah commands believers to "*make not mischief* (fassad) *on earth.*"[15] The meaning of *fassad* can be interpreted as spoiling the natural functioning of the world or spoiling or degrading of natural resources (Amery, this volume). The Prophet (pbuh) once instructed his companions to return to a bird's nest the eggs they took from it. Islamic scholars and rulers have attached penalties to misuse of water, including polluting or degrading clean water. This opens the door for punishing or fining polluters through modern legislation. Also, the Prophet Mohammad (pbuh) very sensibly forbade urination into stagnant water,[16] and advised to guard against three practices, "evacuating one's bowels near water sources, by the roadside and in the shade" (Al-Sheikh 1996).

The situation in the MENA

Given the strong emphasis on equity in Islam, it is useful to examine the current situation in terms of access to water. The low coverage of water supply and sanitation services in rural areas of less developed countries (LDCs) is well documented. In the poor, arid Muslim countries of the Middle East, the situation is no better. About 20 per cent of the population in developing Muslim countries in the Middle East and North Africa (MENA – that is, Algeria, Egypt, Jordan, Lebanon, Morocco, Palestine, Sudan, Syria, Tunisia, and Yemen) was without access to safe water between 1990 and 1996, and close to 37 per cent was without access to sanitation during the same period (UNDP 1998, table 7).

Because the urban growth rate in the MENA is higher than the overall LDC average,[17] informal settlements in and around cities all over the region are rapidly increasing in size. Few of these urban or peri-urban communities receive water and wastewater services, either because the communities were unplanned or because of legal or political restrictions imposed on public utilities. Many of the community residents rely on informal supplies of water sold by private vendors. In LDCs, on average, such families pay ten to twenty times more per litre of water than the rates paid by residents receiving piped water service – and this can rise to eighty to one hundred times in some municipalities (Bhatia and Falkenmark 1993). A literature search for prices paid by the unserved peri-urban poor in the Middle East revealed almost no data available on the topic. However, during the exceptionally warm summer of 1998 in Jordan, the city of Amman suffered a severe water shortage, exacerbated by an odour problem. The public was forced to buy water from vendors, and the black-market price of water delivered by truck tankers reached up to US$14 per cubic metre (Bino and Al-Beiruti 1998). Even under normal weather conditions in Jordan, some of the poor pay a very high price. An informal survey (conducted during an IDRC trip to Amman in December 1998) found that in the Al Hussein refugee camp in Amman, residents not connected to the municipal water system were buying water from their connected neighbours for prices ranging up to US$2 per cubic metre – four times the rate paid by the served customers, whose tariff includes sanitation. Under most conditions, US$2 per cubic metre is greater than the maximum theoretical price for municipal water service as measured by the cost of desalinizing seawater and distributing it.

The issue of water and equity in the MENA requires more investigation, based upon methodical, formal studies. Because the unserved poor live in informal, often unpleasant, forgotten settlements, they are often ignored by mainstream researchers. However, there is no reason to believe that the prices paid by the unserved peri-urban poor are any less in the MENA than in those countries for which information is available. Clearly the current situation is inequitable, and the primary water right under Islam – *haq al shafa* (the right to quench thirst) – is being compromised.

Water demand management

This section discusses both non-economic and economic approaches to water demand management (WDM) within the context of Islam. The non-economic approaches discussed at the workshop included water conservation and wastewater reuse. While not discussed at the workshop, Islam and family planning is also briefly discussed in this section.

Non-economic instruments

Water conservation

The Quran makes two clear statements regarding water that support water demand management. First, the supply of water is fixed, and second, it should not be wasted. The statement that water supply is fixed, and that therefore, at some point, demand must be managed because supplies cannot be infinitely increased is: "*And we send down water from the sky in fixed measure.*"[18] The Quran then tells humans that they may use God's gifts for their sustenance in moderation, provided that they commit no excess therein: "*O Children of Adam!... Eat and drink: But waste not by excess, for God loveth not the wasters.*"[19]

The *hadith* are even more explicit. The Prophet Mohammad (pbuh) "used to perform ablution with one *mudd* of water [equal to $\frac{2}{3}$ litre] and used to take a bath with one *sa'* up to five *mudds* [equal to 2–$3\frac{1}{2}$ litres]."[20] This *hadith* demonstrates the logical approach to sustainable water use in arid Arabia where the Prophet lived. However, the Prophet forbade waste even in conditions of seeming plenty when he said "Do not waste water even if performing ablution on the bank of a fast-flowing (large) river."[21]

Given the clarity of these examples, it is surprising that they are not used more widely to promote water conservation in predominantly Muslim countries. As noted by Atallah (this volume), ordinary Muslims support the idea of being educated about the environment by their religious leaders. In a 1993 survey in Jordan, 64 per cent of the respondents believed that the imams should play an important role in environmental education and public awareness, but only 34 per cent felt that imams were already doing so.

However, policy-makers are beginning to appreciate the value of including religious and cultural values in public awareness and education strategies. Falkenmark (1998) noted recently that regardless of people's culture or religion, "spirituality and ethics are very important for influencing behaviour." Islamic teachings on water conservation are beginning to be incorporated in WDM strategies in predominantly Muslim countries. In Afghanistan, the World Health Organisation (WHO) launched a health education programme through mosques. The programme included training imams on proper health practices, water conservation, and the importance of safe water, proper sanitation, and hygiene in the prevention of diseases. The imams then prepared and gave *khutba* (sermons) on the topic during the congregational Friday prayer (see Atallah, this volume). In Jordan, imams from mosques in the Amman Governorate were educated on water scarcity in the country and the need for public co-operation to address it in a joint programme of the Ministry of Water Resources and the Ministry of *Awqaf* and Islamic Affairs.

Because little information is available to evaluate the effect of these programmes, further research is needed on the topic. However, Shah's chapter discusses a notable exception. He describes a pilot project in Pakistan in a small town, Dijkot, and its surrounding area. The aim was to overcome the shortage of water for domestic uses in the town and for irrigation in the surrounding area. In both cases, the users at the beginning of the water distribution system (the head of the canal in the irrigation district) were taking more than their fair share of water by installing illegal pumps and pumping directly out of the system. An informal group ran a publicity campaign with the participation of the imams in the local mosques and students at the town's religious school. The main message, delivered by the imams on Fridays, and during daily discussions in the mosques, was that "taking another person's share (of water) was a sin and morally wrong." The results were surprising – the number of complaints registered about lack of water decreased by 32 per cent in the town and by 26 per cent in the irrigation district. These results show that in rural areas of Pakistan at least, where the local *Maulvis* – an honorific title for local Muslim leaders or imams in India and Pakistan – enjoy considerable respect and following among the people, focusing on religious values can be surprisingly effective.

In general, according to both Shah and Atallah (this volume), public awareness programmes need to be holistic and multidisciplinary. They should not focus solely on mosques or religious schools, but extend to the education system as a whole. Further, and what is rarely the case, programmes should be co-planned by ministries of Education, Water, and Religious Affairs, so as to be multi-disciplinary – with components of applied science, economics, health, and religion. The Egyptian National Community Water Conservation Programme (NCWCP) of 1993–96 concluded "that the strategy of water conservation communication must be global and interactive, and include ... all the actors concerned, such as religious, political, and informal community leaders" (Afifi 1996). Another lesson was that programmes cannot be short, one-time events. Water conservation requires behavioural change at the societal level, which in turn needs careful, long-term plans of action.

Wastewater reuse

The practice of reusing domestic wastewater for irrigation can be traced back more than two thousand years to ancient Greece. Reusing wastewater is an essential component of a demand management strategy because it conserves freshwater for the highest-value uses. However, treating and reusing domestic wastewater has two other advantages: first, reduced environmental effects, and second, enhanced food production and reduced artificial fertilizer use because of the nutrients contained in the wastewater.

Reusing wastewater is not without health risks or obstacles. Raw wastewater is dirty – it looks and smells bad – and, more importantly, it contains pathogens, including bacteria, viruses, and helminths (parasitic worms), that can cause illness or even death. Given the importance of cleanliness in Islam,[22] and that many MENA countries have minimal wastewater treatment, it is common to hear Muslims declare that wastewater reuse is undesirable, or even *haraam* (unlawful according to Islam). However, as Abderrahman's illuminating case study of Saudi Arabia outlines, reusing wastewater is not *haraam*, provided that it will not cause harm. After a detailed study, in consultation with scientists and engineers, the Council of Leading Islamic Scholars (CLIS) in Saudi Arabia concluded in a special *fatwa* in 1978 that treated wastewater can theoretically be used even for *wudu* and drinking, provided that it presents no health risk (CLIS 1978).

Except in space travel, it is neither cost-effective nor necessary to treat wastewater to such an extent that it achieves a quality necessary for drinking, and the Saudi Arabian scholars did not encourage this practice under normal circumstances. However, treated wastewater can certainly be reused in irrigation, following the WHO guidelines (Mara and Cairncross 1989) devised to protect human health. The guidelines divide irrigation into two main categories – restricted or non-restricted. The necessary wastewater quality, defined in terms of faecal coliform and helminth egg levels, varies depending upon the intended use of the wastewater. Wastewater used for unrestricted irrigation requires more thorough treatment because it can come into contact with edible crops grown at ground level, as well as in sports fields and public parks. Wastewater for restricted irrigation (i.e., of fruit trees, pasture, and fodder crops) requires less treatment, because these can be irrigated with water of lower quality without posing a threat to human or animal health.

On the basis of the 1978 *fatwa*, wastewater reuse in Saudi Arabia expanded greatly. In 1995, the kingdom reused about 15 per cent of its treated wastewater for irrigating date palms and fodder, such as alfalfa. Moreover, ablution water at the two holy mosques in Mecca and Medina is recycled for toilet flushing, thus conserving expensive desalinized seawater. In Kuwait, more than seventeen hundred hectares of alfalfa, garlic, onions, aubergines, and peppers (following the WHO's guidelines) are irrigated using treated wastewater. In Jordan in 1998, 70 million cubic metres (MCM) of treated domestic wastewater were reused. All of this was used for restricted irrigation and accounted for 12 per cent of all water used for irrigation in Jordan (Ministry of Water and Irrigation, Jordan, 1998).

Al-Khateeb's chapter contrasts Islamic precepts regarding wastewater reuse with the actual sociocultural context in Palestine. Almost all the surveyed farmers believed that wastewater reuse was allowable in Islam provided that the practice would not be harmful, and they noted the advantage of irrigating with treated wastewater, which contains valuable nutrients. Most significantly, the farmers were willing to pay up to US$0.24 per cubic metre to buy treated wastewater, and 67 per cent of consumers surveyed were willing to buy crops irrigated with this type of water. The willingness of consumers to buy such products is likely to rise if they are educated by studies such as the one carried out by Al-Khateeb. Wastewater from two pilot secondary treatment plants was used to irrigate eggplants, peppers, apples, grapes, and peaches, and the washwater and the flesh of the fruits and vegetables were tested in a laboratory at the Palestinian Ministry of Water Resources. It was concluded that all of the food was safe to eat. This study supports the WHO's contention that its own guidelines can be relaxed when vegetables such as eggplants and peppers are eaten cooked.

Replacing freshwater with treated wastewater for agriculture will not be easy. Some plants, such as citrus fruits, cannot withstand the high salinity levels in domestic wastewater, but perhaps they should not be grown where freshwater is scarce (see the discussion on food security below). The areas for wastewater irrigation must be carefully selected to avoid contaminating shallow aquifers overlain by permeable soils. However, given that people in the Middle East are already frugal in their water use, and that freshwater will increasingly be taken away from agriculture, expanding wastewater reuse in agriculture is one of the most important WDM policy initiatives in the MENA. Also, because safe reuse depends on adequate treatment, it is vital that virtually every drop of wastewater receive at least some treatment in the region.

Providing effective wastewater treatment has proven to be a challenge in most MENA countries because centralized, mechanical wastewater treatment plants are often unaffordable and, for various reasons, cease to operate efficiently after some time. Perhaps even more importantly, they are designed with wastewater disposal, not reuse, in mind. Most MENA countries will have to implement decentralized, low-cost, natural waste treatment systems, for reuse on or near site. Researchers supported by IDRC are currently pilot-testing grey-water treatment using on-site, small-scale trickling filters for home gardens in the low-density hill settlements surrounding Jerusalem, aquatic wetlands using water lettuce or duckweed in the Jordan Valley and Morocco, and low–mechanical-content activated sludge in Egypt.[23]

Family planning

As noted in the introduction, per capita water availability in the MENA dropped from 3,300 in 1960 to 1,250 m^3/p/y in 1996, and is expected to decline to 725 m^3/p/y by 2025. The main reason for this decline is the population explosion in the MENA – from 92 million in 1960 to about 300 million in 1999. Population in the region will pass the half-billion mark by 2025.

Family planning will not reduce the average water consumption of a given population; however, it can help prevent further reductions in overall availability of water per capita. Consequently, it can be considered a WDM tool – in many countries, without family planning, other strategies to manage water demand will have little or no effect. For instance, with a 1997 per capita water availability of 255 m^3/p/y, Yemen is already terribly short of water. Yet its 1997 population of 16.1 million is expected to nearly double within 20 years (World Bank 1999, table 2.1), which will nullify the impact of ongoing measures to manage water demand, and exacerbate an already desperate situation.

Thus, it is reasonable to consider whether family planning is allowable in Islam and, if so, whether it should be encouraged. Although family planning was not a topic at the workshop, it is discussed briefly here.

A few Islamic scholars believe that birth control is not allowable in Islam, because the Prophet (pbuh) encouraged large families with the words "Marry women who are loving and very prolific."[24]

However, while the Quran makes it clear that children are a blessing from God, it also cautions Muslims not to be too concerned with those and other blessings which they covet in this world: "*wealth and children are the allurement of the life of this world; but good deeds are best in the sight of the Lord.*"[25] Still, it is difficult to find evidence in Islam to support a ban on family planning. Although the religion encourages having children, it is not obligatory. Also, companions of the Prophet reported that they practised coitus interruptus, the only contraceptive method known at the time. Although the Prophet discouraged coitus interruptus in his time,[26] he did not forbid it. Because family planning is not prohibited in either the Quran or the *hadith*, the large majority of scholars believe that, in principle, contraception is allowable in Islam. However, a few simple conditions apply. First, family planning or contraception is only permissible within the Islamic definition of a familial relationship between a man and a woman – that is, the couple must be married. Second, contraception should have the mutual consent of the couple, according to the Prophet's saying: "A man must not practise withdrawal (coitus interruptus) without his wife's consent."[27] Contraception also cannot be imposed upon the couple (Hathout 1989, 228). Finally, the chosen method must truly control conception, rather than birth – that is, it cannot act by causing an abortion.

If these conditions are satisfied, most jurists believe that, following the principle of *maslaha*, if it is in the genuine interest of a society for people to practice family planning, the government should encourage it. For instance, in 1964, the rector of Al Azhar University in Cairo issued a *fatwa* on the acceptability of family planning, noting that "greater numbers were only required in ancient days so that Islam would survive" (Peterson 1999).

Given the support of most scholars, many predominately Muslim countries such as Algeria, Egypt, Iran, Morocco, and Tunisia have a definite government policy on population, while others encourage NGOs to distribute contraceptives and disseminate family planning knowledge: for example, Iraq, Jordan, Sudan, and Syria (Hathout 1989, 225). In fact, according to the UN, Iran, which initially encouraged population growth after the revolution, has emerged as a model of family planning. Beginning in 1987, to counter overcrowding, housing shortages, pollution, and unemployment, the government launched a major family planning program. The minimum age for marriage was increased, and every Iranian couple must attend mandatory classes on birth control before even applying for a marriage licence (Wright 2000, 133). All forms of contraception are free. The dramatic result, a halving of the population growth rate to less than 1.47 per cent in less than a decade, has earned Iran the UN Population Award for 1999 (Peterson 1999). The total number of births per woman (fertility rate) dropped from 6.7 in 1980 to 2.8 in 1997. Efforts in other predominately Muslim countries have also been very successful. Over the same period, the fertility rate dropped from 5.1 to 3.2 in Egypt, from 4.3 to 2.8 in Indonesia, and from 6.1 to 3.2 in Bangladesh. However, women in other predominately Muslim countries, including some of the most water-stressed in the MENA, continue to have very high fertility rates. For instance, the 1997 rate in Yemen was 6.4, while in the West Bank and Gaza it was 6.0 (World Bank 1999, table 2.5).

In summary, family planning is allowable in Islam. Because children are considered a blessing in Islam, family planning should not be encouraged solely for material reasons. Nor would it be allowable for political reasons – that is, to control the population of a particular ethnic or religious group. However, in many countries, continued high population growth is severely stressing existing water resources and the environment. In some of these countries, principles highly valued in Islam, such as equity, quality of life, and the rights of humans and other creatures to sufficient water of good quality, are being compromised. In such cases, where it is in the genuine interest of society to slow down its own growth, family planning should be encouraged. Furthermore, governments should work with religious leaders, because experience has shown that the most successful family programs in predominately Muslim countries have succeeded with the help and support of such leaders. Significant energy devoted to family planning now will lead to fewer social problems, including those related to water, over the next twenty to fifty years, than would otherwise be the case.

Economic instruments

Market approaches to water management, such as increasing tariffs and privatizing utilities, are controversial because water is such a vital social good. Economic measures may be even more controversial in predominantly Muslim nations because of the Islamic precept that water cannot be bought or sold.[28] This section examines economic WDM instruments in terms of water rights and categories, tariffs, privatization, and markets.

Water rights and "ownership" in Islam

In Islam, water is considered a gift from God, so no individual literally owns it. Humans are the stewards of water and other common resources that belong to the community. However, as Djebbar explains (Kadouri et al., this volume), most Islamic scholars have concluded that individuals or groups have the clear right to use, sell, and recover value-added costs of most categories of water. These judgements are based primarily on two *hadith*. First, "It is better ... to go to the woods, [and] cut and sell lumber to feed himself ... than to beg people for help,"[29] which implies that common property resources such as wood and water can be sold and traded (Zouhaili 1992). Second, the earlier cited *hadith* about Othman's purchase of the well at Ruma proves that wells can be owned and traded. Based upon these and other sources, water is categorized in Islam as follows (Sabeq 1981; Zouhaili 1992):

- Private property (water in private containers, treatment plants, distribution systems, and reservoirs). This is water in which work, infrastructure, and knowledge have been invested to obtain it. The "owner" of the "container" has the right to use it, trade it, or sell it.
- Restricted private property (lakes, streams, and springs located in private lands). The owner of the land has special rights over others, but also has certain obligations to them.[30] Within these limits, the owner can trade water like any other good.
- Public property (water in rivers, lakes, glaciers, aquifers, and seas, and from snow and rainfall). Obviously, water in its natural state cannot be bought or sold. However, if infrastructure and knowledge have been invested to withdraw it – for instance, if a public utility constructs a supply, treatment, and distribution system to convey it to people's homes – then the water becomes private property, and the utility has the right to recover its costs. Because of the growing scarcity of water in the MENA, large volumes of freshwater in its natural state are becoming less and less common.

In his time, the Prophet Muhammad (pbuh) discouraged the selling of water, and even "forbade the sale of excess water."[31] Also, as noted, he encouraged Othman to buy the well at Ruma, and give away its water free. These examples reflect the Prophet's desire for the poor and weak to have access to wells controlled by the rich and powerful. It also made sense at the time because water, even though it was relatively scarce, was plentiful enough, clean enough, and accessible enough (through hand-dug wells in shallow aquifers) for sufficient amounts to be available to the very small population of the Arabian peninsula in the seventh century, with almost negligible costs of service provision.

However, it is counter-productive to use this tradition to oppose cost recovery for water services in the current context. In fact, the practice of supplying water (almost) free, under today's conditions of polluted and scarce water supplies has resulted in severe inequities. Subsidizing the collection, treatment, storage, and distribution of water means that increasingly indebted public utilities and governments are able to provide (almost) free water only to the urban rich and middle class. The unserved weak and poor, the very group the Prophet wished to protect, often pay immorally high prices for water in informal markets, or receive water of very poor quality.

Under changing conditions, Muslim leaders can adapt different policies to meet timeless objectives, such as social justice. This point is illustrated by the recent practices of Saudi Arabia, which bases all its laws on *sharia*. Until about twenty years ago, the nation had both ample water and immense wealth, and a small population. Following the Prophet's (and Othman's) example, it provided domestic water nearly free to its citizens. Conditions have changed over the last twenty years, exacerbated by government subsidization of wheat production with cheap irrigation water, which resulted in fossil-water mining. The government has now largely reversed this policy, and the kingdom introduced new water tariffs in 1994, "to acquaint its citizens with the cost of providing water services" (Abderrahman, this volume).

Water tariffs

Evidently, recovering costs for providing water is allowable in Islam. But what is a fair tariff? According to Islam, a fair tariff will lead to greater equity across society. Given the crucial need to conserve water in the region, public awareness and education strategies can only be one element of a multipronged WDM strategy. They must be complemented by economic incentives. Djebbar (this volume) notes that price elasticities of demand in LDCs average −0.45 (higher in rural areas and lower in urban areas), meaning that, all else being equal, a 10 per cent increase in water price will lead to a 4.5 per cent reduction in demand. There is ample room to raise prices for the served middle and high classes. Urban water rates in LDCs are typically less than one-sixth the full cost of water provision (Bronsro 1998). The actual full cost of providing water services will vary from country to country, but in Israel, the only country in the MENA where water is charged at full cost in urban areas, the price (including a surcharge for wastewater treatment) is US$1.00 per cubic metre (Shuval, as cited in Lundqvist and Gleick 1997, 37).

Also, as outlined in Sadr's essay in this volume, full-cost pricing is allowable in Islam. In Iran, where the law is based upon *sharia*, irrigation water must be sold on the basis of average cost, with both operation and maintenance costs and capital depreciation included. This requirement is enshrined in the 1982 Just Distribution of Water Law, the title of which makes the rationale for full-cost pricing self-evident. For urban areas, a 1990 act allows for full (average) cost recovery, including both capital and depreciation costs. As a result of this bill, in 1996 tariffs were increased by 25–30 per cent for household consumption above 45 cubic metres per month, and the tariff for commercial and industrial use was set higher than residential consumption, a step that reversed an earlier policy (see Kazem Sadr's essay in this volume, p. 110).

Where does this leave the poor? In almost every MENA city, a realistic water price, which would allow for reinvestment into the system to serve the unserved poor, would be less than they currently pay, but higher than current prices paid by serviced urban residents.[32] In Jordan, for example, as the informal IDRC survey of December 1998 showed, unserved residents are paying US$2 per cubic metre or more, whereas served residents pay a maximum of US$0.50 per cubic metre, and the full cost of provision is no more than US$1 per cubic metre. Second, tariffs can be structured to supply everyone a lifeline water volume, as is done in Iran, where about the first thirty litres per capita per day (LPCD)[33] are provided free to all domestic customers in urban areas. This approaches the basic water requirement (BWR) standard of fifty LPCD proposed by Lundqvist and Gleick (1997).

Markets and privatization

In Islam, the government may fully recover its costs for providing water to the people. But what about privatization, leading to water being traded like other commodities in the market? First, it is useful to note that a fair and free market finds support in Islam. Muhammad (pbuh) was a businessman prior to his Prophethood, and he set the example for ethical business dealings by earning the title *Al-Amin*, "The Trustworthy," for his personal integrity and fair business dealings. Second, as has been shown, private water rights, separate from land, are allowable for even so precious a commodity as water. Sadr (this volume) notes that in the early Islamic state, as the economy grew, markets for water were established: the first medium of exchange was crops, then the water itself, and finally money.

In a further endorsement of fair markets, the Prophet refused to fix prices for goods in the market, including water, except in special circumstances. In fact, most Muslim scholars agree that a just price for water is that determined by the market, providing that the market is free from unfair practices such as collusion (Khomeini 1989, 4:318–19). This little-known Islamic concept suggests three things. First, a fair price may include not only full cost recovery, but also a fair profit related to the equilibrium price for a good in the market. Second, considering Islam's concern for the protection of the environment, a fair price can also include the cost of treating the wastewater that results from the water's use. Third, privatization is allowable in the water sector. In Iran, municipal water and sewer companies were established under the 1990 Act, which set the legal foundation for private-sector participation in urban water affairs.

Even if full privatization of the water sector is allowable in Islam, that does not mean that it is desirable. Instead, as is generally the consensus in the rest of the world where the private sector participates in providing water services, public-private partnerships are recommended, where the government maintains its "ownership" of water for the community, and allows the private sector to deliver (withdraw, treat, and distribute) water and sewerage services, but regulates the sector to ensure equitable access and also to ensure that quality standards are maintained.

Intersectoral water markets

Faruqui (this volume) argues that enhancing equity means that it is time to take a hard look at how freshwater is allocated in the MENA. Although some water can be saved through domestic conservation practices, the amount is limited because people in MENA already use water very carefully. Rapidly growing populations mean that more water will have to be allocated for domestic purposes.

Where will the water come from? Although the ratio varies from country to country, typically water is allocated in the MENA 10 per cent to industry, 10 per cent to the domestic sector, and 80 per cent to agriculture. Domestic demands are growing and, as MENA countries begin to industrialize, so will the demands of industry – even with recycling – and the water will have to come from agriculture. For instance, Israel's policy is that as urban populations grow, the first priority in water allocation will always be for domestic-urban uses and then industrial needs, followed finally by agriculture (Lundqvist and Gleick 1997). Given the current rate of urbanization, and an unchanging combined urban water consumption rate of 342 LPCD, by 2030, 80 per cent of the fresh water will be used in cities and 20 per cent in agriculture in Israel.

What will be the mechanism of the intersectoral transfer in the MENA? Many recommend allowing the market to reallocate the water. Even with low tariffs, in most cases, the value of water is at least ten times higher in urban areas than it is in agriculture (Gibbons 1986).

What about national food self-sufficiency? An intersectoral transfer policy must be accompanied by increasing urban wastewater treatment, and recycling as much water as possible back to agriculture. Israel plans to reduce its total freshwater volume allocated to agriculture from 70 per cent in 1996 to 20 per cent by 2030.[34] This will be accompanied by an expansion of wastewater treatment so that 80 per cent of urban wastewater will be treated and recycled back to agriculture. This will leave the country with essentially the same amount of water for agriculture as it has at present.

The hard reality is that most MENA countries simply do not have sufficient water for national food self-sufficiency, so this concept must give way to one of national food security (Lundqvist and Gleick 1997, 22), or regional food self-sufficiency, and imports of "virtual water" through the purchase of foods and products produced where it is most efficient. In addition to Israel, water-poor countries such as Botswana have already accepted this fact; and the latter does not have a policy of food self-sufficiency but tries to ensure food security by annual negotiations with suppliers of cereals. Shuval (cited in Lundqvist and Gleick 1997) suggests that a small amount of freshwater, 25 m^3/p/y, should be reserved for domestic production of fresh vegetables, which have high economic and nutritional value. Some of this production may be met by the growing practice of urban agriculture – intensive vegetable production may use as little as 20 per cent of the water, and 17 per cent of the land, used by rural, tractor-cultivated crops (UNDP 1996). Such urban garden vegetables will usually be cheaper for the poor than imported ones. Where feasible, most other crops in arid countries will have to be grown, increasingly and eventually solely, with treated wastewater.

Regulated water markets have been successful in developed countries such as Chile and the United States. In 1991, during a drought period, the California Water Bank purchased water from farmers for about US$0.10 per cubic metre, representing 25 per cent more benefit than the value of the water if used for planting a crop. The water was then sold at an average price of US$0.14 per cubic metre to supply critical urban and agricultural uses (Bhattia and Falkenmark, 1993). In Jordan, the government paid farmers US$120 per hectare for not planting vegetables and annual crops in 1991, a clear case of trading established water rights (Shatanawi and Al-Jayousi 1995).

Are intersectoral water markets allowable in Islam? Two main prerequisites of water markets are that clear rights must exist to water separate from land and that those rights must be tradeable. As already discussed, according to *sharia*, for most categories of water, these prerequisites exist. However, is intersectoral reallocation desirable from an Islamic viewpoint? The priority of use in Islam has been presented, and irrigation has third priority. Obviously, as a population evolves from rural and agrarian to urban and industrial, reallocation is not only permissible, but is required to preserve equity, and the primacy of the right to quench thirst.

In fact, intersectoral transfers through water markets are inevitable.

Already, the growing scarcity of water and its high black-market price have resulted in unregulated water markets all over the MENA, including in Jordan and Palestine. Unregulated markets without necessary legal, institutional, and economic measures in place can lead to unsustainable practices such as in India, where groundwater tables have dropped alarmingly as a result of farmers selling their water to other farmers or cities – ironically, by pumping with subsidized energy.

Governments need to set a vision for national water allocation, and regulate the sector, so that transfers will be slow, constant, and thoughtful. Using the above values, if we assume that one hundred units of renewable water are available to a country as a whole, transferring eight units from agriculture requires only a 10 per cent increase in sectoral efficiency, but this nearly doubles the amount available for domestic purposes: and this is not considering that the same volume, in treated wastewater, may be returned to irrigation, where feasible. In fact, demand management in rural areas is far more likely if users have an economic incentive to voluntarily trade their water use rights. Also, it has been proven that it is possible not only to maintain agricultural production, but even to increase it while reducing the use of water, especially when beginning with the low-efficiency irrigation practices common in most MENA countries. In Africa, for example, in Kenya (Machakos) and Niger (Keita), increases in agricultural production have been achieved while reducing the use of water or reversing land degradation (Templeton and Scherr 1997).

However, in contrast to the wave of neoclassical economics engulfing the world, which at times becomes almost a religion in itself, the rationale for the reallocation is not economic, it is social – the desire to enhance equity. A market approach is merely a tool that a government can use to increase fairness in its society. If regulated water markets are to be used as a tool by MENA governments, then they must put in place legal, institutional, and regulatory mechanisms to ensure that the markets operate fairly and efficiently. Primary among these are institutional mechanisms that will allow for community input and participation in the process (see next section) so that the hard choices necessary for equitable allocative efficiency are made by everyone concerned. However, most developing countries do not yet have the legal, institutional, regulatory, and economic prerequisites to establish sustainable and equitable water markets.

Integrated water management

Biswas set the context for a discussion on integrated water management at the workshop. The 1992 United Nations Conference on Environment and Development in Rio de Janeiro affirmed that "The holistic management of fresh water as a finite and vulnerable resource, and the integration of sectoral water plans and programmes within the framework of national economic and social policy, are of paramount importance for actions in the 1990s and beyond."[35] Simply put, integrated water management should address all water resource management issues in relation to each other, and to the water sector as a whole, with the ultimate goal of promoting equity, efficiency, and sustainability. Because the water resource sector has many vertical and horizontal linkages, such a system cannot exist without an integrated approach that can determine both micro- and macro-level effects of decisions and practices carried out throughout the sector. Biswas highlights some elements of water resource management that need to be integrated. These include:

- Water quality and quantity;
- Technical, environmental, and social issues;
- Land and water uses;
- River basin, estuarine, and coastal management;
- Legal frameworks (that is, cohesive systems of laws and policies); and
- Community-based, national and international water resource management.

Some attempt has been made to establish an Integrated Water Resources Management (IWRM) framework. For example, Agenda 21, Chapter 18 of the Rio de Janeiro conference lists a number of IWRM programme activities.[36] This is a good first step, but developing an integrated water management framework is one thing and actually practising it is far more difficult, even for developed countries. Far more research and additional pilot projects are required to identify how these components should be integrated and who will integrate them.

The workshop was able to touch upon some aspects of Islam and integrated water management by looking at three levels of management: local, national, and international.

Community-based water management

Development practitioners and policy-makers are beginning to accept the principle that water management should be decentralized and that priorities should be set and decisions made at the lowest appropriate level. In many cases, participatory approaches in which local communities help plan, pay for, implement, and run projects that concern them are more likely to be sustainable. For instance, in Uganda, a national policy of decentralization and user-pay has doubled the coverage of water supply from 18 per cent in the 1980s to 36 per cent in 1996, and coverage for sanitation has risen from 20 per cent to 45 per cent during the same period (Lundqvist and Gleick 1997). IDRC is currently supporting a study in India and Nepal that examines the use of local strategies for WDM and conservation to increase water supply and sanitation coverage.[37]

Almost no abstracts were submitted for the workshop on the Islamic perspective relating to decentralization and community participation, and it is evident that further research is necessary in what is a very complex topic. However, based on an informal presentation made by Saeeda Khan of IDRC to initiate debate, and the discussion that followed, consensus among workshop participants was reached on four main points.

First, in contrast to the centralized decision-making system in many Muslim countries, the input of the community on any matter that concerns it, including water management, is mandatory in Islam. In the Quran, believers are defined as those who, among other things, "*(conduct) their affairs by mutual consultation....*"[38] This approach is required by all leaders in Muslim countries and was even required of and followed by the Prophet Muhammad (pbuh) himself.

Second, according to Islam, this consultation is required of all those who are entitled to a voice,[39] including women. In addition, because women are mainly responsible for collecting water in developing countries and are consistently more concerned about the related issues of hygiene and waste management, their input is as important as or more important than that of men. Yet women in most developing countries – regardless of religion or culture – have historically been left out of decision-making. Certainly, in most Muslim countries, despite the rights accorded to them by Islam fourteen hundred years ago, and the Prophet's consultation with his wives and other women,[40] decision makers, who are almost all men, often do not follow the example set by him.

Third, both true community participation and Islam require communities and individuals to be proactive. As noted by Lundqvist (1997), "users must assume responsibilities (including paying fair prices) alongside rights and benefits." One cannot simply sit back and complain that the government has not provided water or wastewater services. Social responsibility begins with individuals and Muslims must help themselves and their communities, as enjoined by the following *hadith*: "No doubt, one had better take a rope (and cut) and tie a bundle of wood and sell it ... rather than ask another who may give him or not."[41]

Fourth, because equitable water management ultimately depends upon a concern for fairness at the individual level, this change necessarily has to happen at the grassroots level. People learn from those nearest them, and individuals in positions of respect, whether for their values or the level of their education, carry a responsibility to propagate concepts such as equity, conservation, environmental protection, and self-help, and to act on them within their own communities. Because these concepts are neither exclusively religious nor exclusively secular, educated religious leaders carry an added responsibility because their knowledge of these issues is reinforced by both their religious and secular knowledge. Many successful community development projects can be attributed to the pro-active leadership and examples shown by educated individuals with strong values such as Mother Theresa in the slums of Calcutta or Dr. Akthar Hameed Khan of the Orangi Pilot Project in Karachi, who inspired and motivated the community and led by example.[42]

National-level water management

No abstracts were submitted for the workshop on the Islamic perspective on national-level water management. Furthermore, in general, little work has been done on the integration of local, regional, and national-level water management. However, a few points can be made here.

If Islamic water management is distilled into one principle, it is water management that balances equity for all of God's creatures as a whole. A nation-state cannot balance equity, efficiency, and sustainability across society without taking a holistic approach that acknowledges the interdependence of water issues.

Principles such as equitable tariffs for the society, protection of the environment, and food security require a discussion and integration of technical, environmental, economic, and social policies that must be sustained by grassroots input, but must ultimately be discussed at national level. The effects of increasing urban water prices and perhaps even irrigation prices can only be analysed at a national level because some effects, in the short term at least and perhaps in the longer term, will be negative, but others will be positive. For instance, raising water prices for those who can afford them may make it possible to serve the unserved poor who currently pay very high prices, and lead to greater equity for the society as a whole – nicely put by the Global Water Partnership slogan, "Some for all, instead of all for some."

The *hadith* "Do not commit any harm or injury to yourself, and do not cause harm or injury to others"[43] and those outlined in the preceding section on rights of the environment, collectively instruct Muslims not to conduct acts that will harm themselves, other creatures, or the environment. This principle cannot be properly upheld if a nation-state does not have a monitoring system to gauge harm to all creatures and the environment. This requires the integration of social, economic, and environmental policies and the development of laws and enforcement of them to protect land and water resources. It also implies the need for environmental, social, and health-impact assessments.

Water reallocation requires that hard decisions be made by all those concerned so as to make fair choices. The inevitable choice of moving from a policy of food self-sufficiency to food security necessarily requires an integrated set of policies and discussions among departments of trade, tourism, industry, water, and agriculture. Because states must be able to earn enough foreign currency from industrial exports and tourism to purchase food produced elsewhere in the world, they must have stable trading relationships (implying just peace between neighbours, see below), and the political situation must be such that food cannot be withheld for political purposes. Also, reallocating water from agriculture to urban areas will leave some farmers jobless, and alternative employment strategies and social safety nets must be considered at the national level by various ministries.

International water resource management

Ultimately, water management principles must guide interactions not simply between individuals, but also between sovereign states, because water does not follow national boundaries. In the Middle East, for instance, the Nile basin is shared by ten countries and the Rum Aquifer is shared by Jordan and Saudi Arabia. Little has been written about Islam and international water management, but Hussein and Al-Jayyousi's essay in this volume explores the topic and offers some preliminary conclusions. Internationally, the latest consensus in global water management is reflected by the thirty-three articles drafted by the International Law Commission and approved by the United Nations General Assembly in 1997. The convention is now awaiting ratification by member states. Its four most important principles are:

- Equitable and reasonable utilization of international rivers (Article 5);
- Avoidance of significant harm and compensation (Article 7);
- Cooperation among riparian states (Article 8); and
- Protection and preservation of international rivers and associated ecosystems (Articles 5, 8, 20, and 21).

These international law principles are in harmony with Islam because they are based on universal values. These values are embodied in the Islamic concepts that water is a gift from God to his creatures and hence that all of creatures have the right to use water to quench their thirst, that water should be apportioned equitably for other uses, and that no one has the right to withhold surplus water from others. A further concept, of avoiding significant harm to others, is emphasized by the *hadith* concerning committing harm or injury to oneself and to others as well as by another *hadith*, "He who eats to his fill while his neighbour goes without food is not a believer,"[44] which is applicable to drink as well as food. "Neighbour" can be considered as an individual or as a neighbouring state, whether Muslim or of another faith. Also, if harm does occur, according to *sharia*, it carries a liability – that is, the one against whom it is committed must be compensated. In addition, relevant universal values are embodied in the Islamic requirement of *shura* (consultation on all matters of mutual interest), as well as in the emphasis in Islam on protecting and preserving water and its ecosystems by avoiding *fassad* (mischief or harm).

In practice, however, the convention, even if ratified by all UN member states, will only be an unenforceable guideline, and there are currently many international water-sharing disputes where nation-states are not following these principles. For example, the per capita water usage in Israel is about 330 LPCD, whereas in Palestine it is about 50 LPCD. If there is to be just peace in the region, Israel and Palestine will have to co-manage the mountain aquifer underlying Israel and the West Bank, and share its water equitably, and the IDRC is currently supporting a research project on the joint management of the aquifer.[45] Similarly, Iraq, Syria, and Turkey must work out an equitable agreement to apportion their shared waters. Because the international water management principles are strongly and explicitly supported by Islam, some workshop participants suggested that predominantly Muslim nations should take water-sharing disputes to an Islamic council authorized to mediate and judge on disputes. Although negotiations for equitable water sharing between states are difficult, they are not impossible, particularly when mediated – as was shown by the 1960 Indus Basin River Treaty between India and Pakistan, brokered by the World Bank, which prevented war between the two countries. Islamic *sharia* provides legal standing to any contract or obligation that has been made between two parties and makes this contract binding.

Islamic water management principles

The workshop participants reached a consensus on Islamic water management principles under the following headings: water as a social good; water demand management; and integrated water resources management. The overriding principle under all three is that of ensuring equity.

Water as a social good

- Water is first and foremost a social good in Islam – a gift from God and a part of, and necessary for, sustaining all life.
- Water belongs to the community as a whole – no individual literally owns water.

- The first priority for water use is access to drinking water of acceptable quantity and quality to sustain human life, and every human being has the right to this basic water requirement.
- The second and third priorities for water are for domestic animals and for irrigation.
- Humankind is the steward of water on earth.
- The environment (both flora and fauna) has a very strong and legitimate right to water and it is vital to protect the environment by minimizing pollution. Individuals, organizations, and states are liable for harm that they have caused to the environment or to the environmental rights of others, including water use rights.
- Water resources must be managed and used in a sustainable way.
- Sustainable and equitable water management ultimately depends upon following universal values such as fairness, equity, and concern for others.

Water demand management

- Water conservation is central to Islam. Mosques, religious institutes, and religious schools should be used to disseminate this principle so as to complement other religious and secular efforts.
- Wastewater reuse is permissible in Islam; however, the water must meet the required level of treatment to ensure purity and health for its intended purpose.
- Full cost recovery is permissible: that is, the full cost of supplying, treating, storing, and distributing water, as well as the cost of wastewater collection, treatment, and disposal. However, water pricing must be equitable as well as efficient.
- Privatization of water service delivery is permissible in Islam, but the government has a duty to ensure equity in pricing and service.

Integrated water resources management

- Water management requires *shura* (consultation) with all stake-holders.
- All community members, including both men and women, can play an effective role in water management and should be encouraged to do so.
- Communities must be proactive to ensure equitable access to water resources.
- All nation-states have an obligation to share water fairly with other nation-states.
- Integrated water management is a necessary tool to balance equity across sectors and regions.

Recommendations

The workshop's recommendations are directed at varying audiences. The recommendations are not made specifically to the IDRC, the International Water Resources Association (IWRA), or the Inter-Islamic Network on Water Resources Demand and Management (INWRDAM), although some recommendations may be relevant depending upon the mandate of the organization. In some cases, the recommendations are relevant to any water specialists, donor agencies, or policy-makers and, in other cases, they are specific to a Muslim audience.

Water as a social good

- Cooperation and sharing of knowledge of water resource management should be encouraged among Muslim scientists and countries by developing a network to promote equity.
- For the same purpose, cooperation and sharing of knowledge of water resource management should also be encouraged among scientists and countries regardless of religion.

Water demand management

- Non-economic incentives for conserving water, as well as penalties for wasting it, should be identified.
- Wastewater should be properly treated and reused.

Integrated water resource management

- Muslim countries need to agree upon the mandates of various existing international Islamic organizations, empower them to rule on conflicts over water use rights between Muslim states, and abide by their decisions.
- In disputes between Muslim states and those of other faiths, all parties should comply with fair and just rulings by appropriate international organizations.

Further research

The workshop's recommendations for applied research projects or studies address questions left unanswered by the workshop, specific gaps in knowledge identified at the workshop, and proposals made by others to realize concrete benefits from new insights at the workshop. Although the suggestions were discussed at the workshop in detail, they have been left fairly general here, to allow interested parties to identify specific objectives and elements.

The suggestions are made particularly to policy-makers and to donor agencies, but the audience for each suggestion depends upon its nature. For instance, applied research particularly concerned with Islamic issues may be beyond the mandate of such organizations as the IDRC or the IWRA, and may be more relevant to such organizations as the Islamic Development Bank (IDB) and INWRDAM.

Water as a social good

- Conduct rigorous scientific surveys of equity of access to water and sanitation, by identifying the volume, quality, and price paid in the MENA by both the unserved poor in the informal sector and the served middle- and high-income classes. The surveys should capture such information as price paid per capita, percentage of income spent on water, and willingness to pay.
- Investigate the priority of rights to water relevant to current economic, demographic, and settlement patterns, in particular clarifying the rights of the environment and the right of wild animals and flora to water.

Water demand management

- Conduct a wide-ranging pilot study to integrate religious elements into a comprehensive programme of public education and awareness projects to encourage conservation and reuse, with particular emphasis on women and girls, who are often left out of such programmes because their religious learning does not occur in mosques or schools.
- Examine water tariffs, including the elasticity of water demand in different sectors and under different conditions, willingness to pay for improved water quality, tariff structures, and modalities of subsidies (on water, income, with stamps, and so on) for the poor.
- Investigate how intersectoral water reallocation using markets may be carried out more equitably by examining such issues as:
  - Effect of unregulated markets;
  - Development of models to analyse the social, environmental, and economic effects of intersectoral reallocation;
  - Farmers' willingness to sell freshwater use rights to the domestic and industrial sector in exchange for treated wastewater;
  - Methods for monitoring third-party effects;
  - Institutions that could serve as an interface between buyers and sellers; and

Legal reforms and private and state ownership of surface and groundwater rights.

- Explore methods to improve the efficiency and equity of water use in rural areas, including traditional and indigenous practices and technology.
- Use pilot decentralized, community-run, low-cost, real-world-scale wastewater treatment and reuse projects in a variety of conditions to methodically investigate how to make such projects sustainable.

Integrated water resource management

- Analyse models of community-based water management and stakeholder participation:
  - Identify contemporary and historical case studies (successes and failures) of community management in the Muslim world and regions of other faiths and develop models for dissemination;
  - Assess how to move beyond simply involving communities and water users' associations in decision-making and to empower them;
  - Explore how to develop the common interest between communities; and
  - Develop gender analysis of community-based water management projects in Muslim countries – models for more effectively bringing women into community-based water management.
- Investigate how to take the concept of integrated water management from theory to practice, using various means, for example, by examining successful case studies.
- Research more specific and operational principles of international law, consistent with Islam, including historical practices.

Conclusions

Before it came to mean simply "law," the Arabic word *sharia* denoted the law of water (Mallat 1995). It is, therefore, not surprising that a detailed examination of the Quran and the *hadith* shows that Islam makes a remarkable number of specific statements about water management.

There is no contradiction between what Islam says about water management and the emerging international consensus on the issue, as reflected by recent accords such as the Dublin Principles or the UN Water Convention. In fact, the Islamic water management principles are not unique. Some of the same principles could be derived by studying other faiths, their holy books, and the lives of their prophets. As one delves into Islam, one encounters values common not only to the other two Abrahamic religions, Christianity and Judaism, but also to many other world-views and religions. But clean water has always been scarce in the Middle East where Islam emerged and where for many centuries most Muslims lived, whereas water has only recently begun to become scarce in regions such as Europe where for many centuries the majority of Christians lived. Hence, the rules governing water management are probably more specific and detailed in Islam than in most other religions.

The principles, recommendations, and suggested further work outlined at the workshop were noted in the previous section. The most important findings are as follows.

- Water is a social good owned by the community. Provided that equity is maintained, as in Iran where all urban residents receive a lifeline volume of water free to meet their basic requirements, Islam allows for private sector involvement in service delivery, and up-to-full-cost recovery for water and wastewater treatment services.
- In contrast to the current situation in the MENA, the priority of water rights is first, domestic uses; second, livestock watering; and third, irrigation. The environment has very strong and specific water rights, and individuals, organizations, and states are liable for harm that they have caused to the environment, which allows for "polluter pays" legislation.
- As indicated by the *fatwa* and actual practice in Saudi Arabia, wastewater reuse is allowable, and encouraged where necessary, provided that the water is treated to the extent that makes it safe for its intended use.

The workshop suggests that further studies and investigation are required, in areas such as Islam and community-based water management. How Islam, or other belief systems, can be integrated, along with a whole number of other factors, into holistic water management is a larger question.

Further studies of this nature are likely to be beneficial for more effective and equitable water management. The study that led to the *fatwa* in favour of wastewater reuse in Saudi Arabia demonstrated two things: first, *ijtihad*, or innovation, is permissible, relevant, and necessary in today's world; and, second, the specific objectives of Islam, and other religions, are a reflection of the religion's values, such as maintaining equity in society, and are timeless and unchangeable. Some of the means of attaining these objectives, such as mandatory *zakaat*, the charity tax, which is one of the Five Pillars of Islam, are also unchangeable. However, other practices to achieve the objectives, such as reusing treated wastewater to conserve water so that all may share in its benefits, can and must change depending upon specific conditions.

Finally, even though water has always been scarce in the Middle East, per capita availability of clean water has dropped alarmingly only in the past decade, and the rate of decline is accelerating. In other words, up to now, we did not face a water crisis in the Middle East or elsewhere. Muslims, like other people, tend not to react to crises until they are upon them. So the time when Islamic water management principles are likely to become most relevant to Muslims is only now upon us. Important water demand management policy instruments, which must be used to combat this crisis are:

- Encouraging family planning to reduce high birth rates, where appropriate;
- Diverting fresh water from irrigation to domestic and industrial uses;
- Treating domestic and industrial wastewater and reusing it for irrigation;
- Protecting the environment, including legislating and enforcing liability for harm;
- Conserving water in all sectors;
- Exploring private-public partnerships for water services delivery and regulation; and
- Decentralizing water management and managing it at the community level.

These measures all have strong, specific support in Islam, more so perhaps than in other belief systems, which may make it easier to introduce such policies, if they are accompanied by comprehensive public awareness programmes, including religious elements.

Notes

1. 47:12.
2. 16:65.
3. 21:30.
4. Muslim 1411.
5. Al-Bukhari 1.12.
6. Al-Bukhari 3.838.
7. 59:7.
8. Abu-Dawood 3470.
9. 6:38.
10. Al-Bukhari 8.38.
11. Al-Bukhari 5550, in *Hadith Encyclopedia*.
12. Al Bukhari 4.538.
13. 6:99.
14. 35:77.
15. 2:11.
16. Muslim 553.

17. For the period covering 1995–2015, the average urban growth rate for LDCs is 2.9 per cent, in contrast to 3.2 per cent for the MENA countries in which IDRC supports projects: Algeria, Egypt, Jordan, Lebanon, Morocco, Palestine, Syria, Sudan, Tunisia, and Yemen.
18. 40:18.
19. 7:31.
20. Al-Bukhari 1.200.
21. Al-Tirmidhi 427.
22. A full discussion of the importance of personal cleanliness in Islam is beyond the scope of this chapter. However, Islam has very specific and detailed rules, easily referenced in the Quran and the *hadith*, for maintaining cleanliness including *wudu*, ablution before prayer; *ghusl*, bath after sexual intercourse and before prayer; trimming body hair from underarms and intimate areas; and proper washing, with water, after defecation.
23. The type of treatment will vary, depending on specific local conditions such as type of soil, land availability, and the intended end use of the wastewater. For more information on IDRC-supported applied research projects on waste treatment and reuse in the MENA (Egypt, Palestine, Morocco, and Senegal), contact Naser Faruqui at IDRC's Cities Feeding People Program.
24. Abu-Dawood 2045.
25. 18:46.
26. See, e.g., Al Muwatta 29. 95–100, Al-Bukhari 7. 135–37.
27. Abu Dawood, cited in Hathout 1989, 227.
28. The principle "water is an economic good" was worded in a very general way at the 1992 UNEP Dublin Water Conference because, among other reasons, some participants from predominately Muslim countries argued that selling water was against Islam (A. Biswas, personal communication).
29. Muslim 1727.
30. For instance, one has the right to trespass on private lands to satisfy thirst if one's life or health is threatened, and no one has the right to hold back surplus water (Al-Bukhari 9. 92).
31. Muslim 3798.
32. In Ivory Coast, in 1974, only 30 per cent of the urban population and 10 per cent of the rural population had access to safe water. By 1989, 72 per cent of the urban population and 80 per cent of the rural population had access to safe water (water points). This occurred because the private water company Société de Distribution d'Eau de la Côte d'Ivoire was allowed to increase urban tariffs above the level of long-term marginal costs, especially for industrial customers (Bhattia et al. 1995)
33. Five thousand litres per household per month, assuming an average of six persons per household (Sadr, this volume).
34. In fact, the amount of freshwater left over for agriculture may be even less than 20 per cent if Israel eventually allocates some portions of the freshwater currently under its control to its neighbours to achieve a peace agreement (Shuval, as cited in Lundqvist and Gleick 1997, 37).
35. Earth Summit CD-ROM, Agenda 21, Chapter 18, Section 18.6. (IDRC, Ottawa, Canada, FIS No. 92-0608).
36. Ibid., Section 18.12.
37. For more information on this project, "Local Strategies for Water Supply and Conservation Management (India, Nepal)," contact David Brooks, at IDRC's People, Land and Water Program.
38. 26:38.

39. One characteristic of Muslims who truly worship and serve Allah is that "conduct in life is open and determined by mutual consultation between those who are entitled to a voice, e.g., in private domestic affairs, as between husband and wife, or other responsible members of the household; in affairs of business, as between partners or parties interested; and in State affairs, as between rulers and ruled, or as between different departments of administration, to preserve the unity of administration" (Yusuf Ali 1977, n. 4578).
40. One example of the Prophet Muhammad's (pbuh) consultation and acceptance of his wives' views is provided by the following *hadith*: "When the writing of the peace treaty (of Al-Hudaibiya) was concluded, Allah's Apostle said to his companions, 'Get up and slaughter your sacrifices and get your heads shaved.' By Allah none of them got up, and the Prophet repeated his order thrice. When none of them got up, he left them and went to Um Salama (his wife) and told her of the people's attitudes towards him. Um Salama said, 'O the Prophet of Allah! Do you want your order to be carried out? Go out and don't say a word to anybody till you have slaughtered your sacrifice and call your barber to shave your head.' So, the Prophet went out and did not talk to anyone of them till he did that, i.e., slaughtered the sacrifice and called his barber who shaved his head. Seeing that, the companions of the Prophet got up, slaughtered their sacrifices, and started shaving the heads of one another" (Al-Bukhari. 3.891).
41. Al-Bukhari 3.561.
42. The Orangi Pilot Project was a highly successful community-based water management initiative of the 1980s. It was designed to provide low-cost sewage disposal and sanitation systems to the low-income settlement of Orangi on the outskirts of Karachi, Pakistan (Hassan 1994).
43. *Hadith* narrated by Said Saad bin Sinan Al Khudri, Al-Baghdadi 1982, 285.
44. Shu'ab Al-Imam-Baihaqui.
45. For more information on this project, contact David Brooks at IDRC's People, Land and Water Program.

REFERENCES

Afifi, Madiha Moustafa (1996), "Egyptian National Community Water Conservation Programme," in *Environmental Communication Strategy and Planning for NGOs, Ma'ain, Jordan, 27–31 May 1996*, Jordan Environment Society, Amman.

Al-Baghdadi, Abu Abd Al Rahman Mohammed bin Hasan (1982), *Jamma Al Anloum Wal Hiram* [Collection of the sciences and wisdom], 5th ed., Dar Al Manhal, Cairo.

Al-Sheikh, Abdul Fattah al-Husseini (1996), *The Right Path to Health – Health Education through Religion: 2. Water and Sanitation in Islam*, WHO Regional Office for the Eastern Mediterranean, Alexandria.

Ansari, M. I. (1994), "Islamic Perspectives on Sustainable Development," *American Journal of Islamic Social Science* 11 (3), pp. 394–402.

Bhattia, R., Cesti, R. and Winpenny, J. (1995), *Water Conservation and Reallocation: Best Practice Cases in Improving Economic Efficiency and Environmental Quality*. World Bank – Overseas Development Institute. Joint Study, Washington, D.C.

Bhattia, R. and Falkenmark, M. (1993), *Water Resources Policies and Urban Poor: Innovative Approaches and Policy Imperatives, Water and Sanitation Currents*, World Bank, UNDP – World Bank Water and Sanitation Programme, Washington, D.C.

Bino, M. J. and Al-Beiruti, Shihab N. (1998), "Inter-Islamic Network on Water Resources Development and Management (INWRDM)," *INWRDM Newsletter* (Amman) 28 (October).

Bronsro, A. (1998), "Pricing Urban Water as a Scarce Resource: Lessons from Cities around the World," in *Proceedings of the CWRA Annual Conference, Victoria, B.C., Canada*, Canadian Water Resources Association, Cambridge, Ont.

CLIS (Council of Leading Islamic Scholars) (1978), "Judgement Regarding Purifying Wastewater: Judgement no. 64 on 25 Shawwal, 1398 AH, thirteenth meeting of the Council of Leading Islamic Scholars (CLIS) during the second half of the Arabic month of Shawwal, 1398 AH (1978)," Taif, Saudi Arabia, *Journal of Islamic Research* 17, pp. 40–41.

Falkenmark, M. (1998), *Willful Neglect of Water: Pollution – A Major Barrier to Overcome*, Stockholm International Water Institute Waterfront, Stockholm.

Gibbons, Diana C. (1986), *The Economic Value of Water*, Resources for the Future, Washington, D.C.

Hassan, Afrif (1994), "Replicating the Low-Cost Sanitation Programme Administered by the Orangi Pilot Project in Karachi, Pakistan," in Ismail Serageldin and Michael A. Cohen (eds.), *The Human Face of the Urban Environment: Proceedings of the Second Annual World Bank Conference on Environmentally Sustainable Development*, World Bank, Washington, D.C.

Hathout, H. (1989), "Ethics and Human Values in Family Planning: Perspectives for the Middle East," in Z. Bankowski, J. Barzelatto, and A. M. Capron (eds.), *Ethics and Human Values in Family Planning: Conference Highlights, Papers and Discussion: XXII CIOMS Conference, Bangkok, Thailand, 19–24 June 1988*, Council for International Organizations of Medical Sciences, Geneva.

Khalid, F. (1996), "Guardians of the Natural Order," *Our Planet* 8 (2), pp. 8–12.

Khomeini, Roohulla (1989), *Ketabul Beia* [The book of choosing a successor], Ismaeilian, Qum, Iran.

Lundqvist, Jan (1997), *Most Worthwhile Use of Water – Efficiency, Equity and Ecologically Sound Use: Pre-requisites for 21st Century Management*, Water Resources 7, Department for Natural Resources and the Environment, Stockholm.

Lundqvist, Jan and Gleick, Peter (1997), *Comprehensive Assessment of the Freshwater Resources of the World: Sustaining Our Waters into the 21st Century*, Stockholm Environment Institute, Stockholm.

Mara, D. and Cairncross, S. (1989), *Guidelines for the Safe Use of Wastewater and Excreta in Agriculture and Aquaculture*, World Health Organization, Geneva.

Mallat, Chibli (1995), "The Quest for Water Use Principles," in M. A. Allah and Mallat Chibli (eds.), *Water in the Middle East*, I. B. Tauris, New York.

Ministry of Water and Irrigation, Jordan (1998), *Yearly Report*. Ministry of Water and Irrigation, Amman.

Peterson, S. (1999), "An Unlikely Model for Family Planning," *Christian Science Monitor*, 19 November.

As-Sayyed, Sabeq (1981), *Fiqh essounna* [Understanding the Prophet's traditions], 3d ed., Dar El-Fiqr, Beirut.

Shatanawi, M. R. and Al-Jayousi, O. (1995), "Evaluating Market-Oriented Water Policies in Jordan: A Comparative Study," *Water International* 20 (2), pp. 88–97.

Templeton, S. R. and Scherr, S. J. (1997), *Population Pressure and the Microeconomy of Land Management in Hills and Mountains of Developing Countries*, Discussion Paper 26. Environment and Production Technology Division, International Food Policy Research Institute, Washington, D.C.

UNDP (United Nations Development Programme) (1996), *Urban Agriculture: Food, Jobs, and Sustainable Cities*, Series for Habitat 2, vol. 1, UNDP, New York.

UNDP (United Nations Development Programme) (1998), *Human Development Report*, UNDP, New York.

World Bank (1999), *World Bank Development Indicators 1999*, World Bank, Washington, D.C.

Wright, R. (2000), "Iran's New Revolution," *Foreign Affairs* 79 (1), pp. 133–45.

Yusuf Ali, A. (1977), *The Holy Qur'an: Text, Translation, and Commentary*, American Trust Publications for The Muslim Student Association of the United States and Canada, Plainfield, Ind.

Zouhaili, O. (1992), *Al-Fiqh wa-dalalatuh* [Islamic jurisprudence and its proof], Dar El-Machariq, Damascus.

2

Islamic water management and the Dublin Statement

Odeh Al-Jayyousi

In the last two decades, the need for new approaches to the assessment, development, and management of freshwater resources has been emphasised at various global meetings. According to the United Nations Development Programme (UNDP 1990), integrated water resources management is based on the perception of water as an integral part of an ecosystem, a natural resource, and a social and economic good. The International Conference on Water and the Environment: Development Issues for the Twenty-First Century, held in Dublin in January 1992, called for new approaches to the assessment, development, and management of freshwater resources (UN 1991; UNEP 1992). Moreover, the United Nations Conference on Environment and Development in Rio de Janeiro in June 1992 confirmed the widespread consensus that the management of water resources needs to be reformed. The conference stated that "The holistic management of freshwater as a finite and vulnerable resource, and the integration of sectoral water plans and programs within the framework of national economic and social policy are of paramount importance for actions in the 1990s and beyond" (World Bank 1993, 24).

Necessary conditions for the success of these approaches include: public awareness campaigns, legislative and institutional changes, technology development, and capacity-building programmes. Underlying all these must be a greater recognition of the interdependence of all peoples, of their norms and values, and of their place in the natural world.

The Islamic perspective toward both man and nature provides a conceptual framework for sustainable resource management. The purpose of this chapter is to compare Islamic water management principles to those enunciated by the Dublin Conference.

The Islamic perspective

Islam covers all aspects of human life. It regulates the relationships between God, humans, and nature. It is based on the recognition of the unity of the Creator and of humans' submission to His will. Muslims believe that everything originates from the One God, and everyone is responsible to Him. Humans are viewed as trustees (*khulafa*) and witnesses (*shahed*). Our role and responsibility is to ensure that all resources, including water, are used in a reasonable, equitable, and sustainable manner.

According to Islam, nature is created by God (Allah) for the benefit of humans. The relationship between humans and nature is based on harmony, because all creatures obey the laws (*sunan*) of God. Humans are urged to explore and use natural resources in a sustainable manner. Muslims believe that by submitting to the will of God peace is brought about and that by harmonizing humans' will with the will of God life becomes responsible and balanced. Every human activity is given a transcendent dimension: it becomes sacred, meaningful, and goal-centred.

The scheme of life that Islam envisages consists of a set of rights and obligations. Broadly speaking, the law of Islam imposes four types of rights and obligations on every person: first, the rights of God; second, his or her own rights upon his or her own self; third, the rights of other people over him or her; and fourth, the rights of those created things that God has empowered humans to use for their benefit.

This chapter focuses on the rights of created things. Muslims believe that God has honoured humans with authority over the countless things that He has created. Everything has been harnessed for us. We have been endowed with the power to subdue them and make them serve our objectives. This superior position gives humans authority over resources, including water. However, this authority must be guided by a sense of responsibility and accountability toward both living creatures and nature. Humans should not waste resources on fruitless ventures nor should they unnecessarily damage them. When humans employ resources in their service, they should employ the best and the least injurious methods of deriving benefit from these resources.

The Dublin principles and Islamic concepts

The Dublin Conference's rationale was that if water and land resources are not managed well, then human health, food security, economic development, and ecosystems will all be at risk. The conference called for fundamental new approaches to the assessment, development, and management of freshwater resources. It stressed the fact that commitment will need to be backed by substantial and immediate investments, public awareness campaigns, legislative and institutional changes, technology development, and capacity-building programmes.

This is fully in line with the basic Islamic perspective underlying Islamic water management (IWM). Many verses in the Quran illustrate the value of water, how was it formed, and its vulnerability: for example: "*We made from water every living thing. Will they not then believe?*"[1] and "*Say: If your stream be some morning lost (in the underground earth), who then can supply you with clear flowing water?*"[2] Principle 1 of the Dublin Statement says: "Fresh water is a finite and vulnerable resource, essential to sustain life, development and the environment." It is evident that Principle 1 is consistent with IWM concepts. Both agree that freshwater resources are limited, vulnerable, and important for life.

The concept of community participation and consensus building is well established in IWM. The Quran urges that decision-making must be based on group consultation and consensus (*shura*). The Quran describes believers as "*those who harden to their lord, establish regular prayer, and who conduct their affairs by mutual consultation.*"[3] The Prophet (pbuh) practised consultation and accepted advice on many occasions. He decided where to camp in Bader, near sources of water, based upon the suggestion of one of his companions, Habbab Ibn Al-Munther (Ibn Hisham 1991, 167–68). This consultation is consistent with Principle 2, which stresses the notion that water management and development should be based on participation of all stake-holders.

The participatory approach involves raising awareness of the importance of water among policy-makers and the general public. It means that decisions are taken at the lowest appropriate level, with full public consultation and involvement of users in the planning and implementation of water projects. This can be done through the establishment of water users associations (WUAs) or other non-governmental organizations (NGOs). They can play a role in adjusting, modifying, or enacting laws and regulations that are consistent with sustainable water management. Islam urges all members of society to take an active and positive attitude toward public concerns. This involvement should be performed through effective communication and consultation.

Each has a social responsibility to conserve water and prevent water pollution. According to Principle 3 of the Dublin Statement, "Women play a central part in the provision, management and safeguarding of water." Likewise, in Islam the responsibility for taking care of resources is not divided by gender. In IWM, both men and women are considered as caretakers of resources.

The role of women in Islam as providers and users of water and guardians of the living environment is well documented. Bringing water from springs and wells was typically carried out by women, and historically, the story of the rituals of pilgrimage in Mecca has been formulated around Hajar, the wife of the Prophet Ibrahim. Her search for water between Safa and Marwa had made these places into sites of remembrance for Muslims. Likewise, Zubaidah, the wife of Al-Rashid, played a key role in building a water canal in Mecca during the Abbasid period. When she performed pilgrimage in 808 A.C. she saw the hardships that pilgrims suffered because of the scarcity of water, and summoned engineers and workers from near and far to construct a canal to transport water from the Ein Hanin spring to Mecca. She was determined to achieve this goal at any cost and she told her finance manager (*khazen*): "Implement it, even if every dig in the ground is worth a dinar" (Hasan 1964). The completed canal was named after her, Ein Zubaidah. This is a clear example of the role of women in Islam in water development and demonstrates how women can show leadership and take social responsibility.

In an Islamic society, both men and woman play a crucial role in making the world a livable place. They act as God's deputies on earth. They both enjoin what is right and forbid what is wrong. A functional division of labour is practised in the Muslim family. The man takes primary responsibility for earning and providing the necessities of life; while the woman takes primary responsibility for managing the household and educating and bringing up the children. As a result, Muslim women can play an important role in conserving water at home and in society. They can convey knowledge, attitudes, and practices that promote conservation, pollution prevention, and sustainable consumption. In the local community or at higher policy levels women can be part of advisory commissions for water planning and management. Instilling values of environmentally sound practices is of crucial significance to the future. Thus, because of their primary role in Islam to educate their children, women have a key position in teaching future generations sustainable consumption patterns to ensure effective use of resources.

Principle 4 of the Dublin Statement states that "Water has an economic value in all its competing uses and should be recognized as an economic good." Similarly, the Prophet Muhammad declared that water should be, together with pasture and fire, the common entitlement of all Muslims. This is why, in many modern Muslim countries, water legislation considers that water resources belong to the whole community, that is, the state or the public domain (Caponera 1992). Based on this notion, public water in its natural state (large lakes and rivers) cannot be sold. Access to water is a right of the community.

Islamic law does, however, distinguish between public and private water. Private water includes that contained in wells, tanks, and other reservoirs. If an additional cost is incurred to convey, treat, and store water, then it is considered to be under private ownership (Zouhaili 1989). This implies that water users have to pay the cost of operation, treatment, and maintenance of water supply systems. However, special consideration must be paid to low-income users who do not have the ability to pay and, for some users, water should be subsidized. In addition, the right to use water can be separated from the land which a watercourse traverses, not by sale but by legacy. Although the water in such a canal is privately owned, everyone has the right to drink from it, but he must not trespass on the land where the canal is situated without the permission of the owner, except in case of necessity. Full private property in water exists only if it is "in custody," that is, in a container. The state has the right to recoup the cost of supplying, treating, and distributing public water.

Conclusions

Islam presents a reference and code of conduct for humans toward resource management. Humans are viewed as trustees (*khulafa*): their role and responsibility is to ensure that all resources, including water, are used in a reasonable, equitable, and sustainable manner. In spirit this agrees with all the principles of the Dublin Statement. Islamic thought agrees that freshwater resources are vulnerable and important for all aspects of life. Participatory approaches (*shura*) in water management should be enhanced at all levels. The role of women in water conservation and awareness is vital in water management, and their role in water education should be enhanced through both formal and informal mechanisms. Research on the reform of the domain of women's role in society is needed. The involvement of women in WUAs and other NGOs must be supported.

Regarding Principle 4, however, that water has economic value, more research should be carried out to clarify the economics of water, water rights, and the value of water. Equity issues in reallocation of water must be addressed from an Islamic perspective. The distinction between public and private water and its implications for water pricing must be explained to the public.

To operationalize IWM principles, a consultative council for sustainable water management and law reform is recommended. This council should be represented by scholars in both science and religion to ensure interdisciplinary learning and help to promote innovation (*ijtihad*). One major task of this council would be to formulate both national and international Islamic water policy. Evaluations of the performance of this council, as well as new rulings (*fatwa*) as they appear, should be accessible to the public.

Notes

1. 21:30.
2. 67:30.
3. 42:38.

REFERENCES

Caponera, D. A. (1992), *Principles of Water Law and Administration: National and International*, Balkema Publishers, Brookfield, Vt.

Hasan, H. I. (1964), *The History of Islam: The First Abbasid era*, vol. 2 (7th ed.), Al-Nahda Library, Cairo.

Ibn Hisham (1991), *The Life of the Prophet*, vol. 3, Dar Al-Jaleel, Beirut.

UNDP (United Nations Development Programme) (1990), *Safe Water 2000*, New York.

UNEP (United Nations Environmental Programme) (1992), *Final Report of the International Conference on Water and the Environment, Dublin*, UNEP, Nairobi.

UN (United Nations) (1991), *A Strategy for the Implementation of the Mar del Planta Plan for the 1990s*, UN Department of Technical Cooperation, New York.

World Bank (1993), *A World Bank Policy Paper – Water Resources Management*, The World Bank, Washington, D.C.

Zouhaili, O. (1989), *Al-Fiqh wa-dalalatuh* [Islamic jurisprudence and its proof], pt. 4, Dar Al-Fikr, Beirut.

3

Islam and the environment

Hussein A. Amery

The purpose of this chapter is to describe the Islamic perspective on natural resource management, with a particular focus on water. Although elements of culture or religion are typically absent from the writings of most academics on natural resource and environmental issues, one culture-aware author states that the word "environment" includes the "biological, physiological, economic and cultural aspects, all linked in the same constantly changing ecological fabric" (de Castro, quoted by Vidart 1978, 471). The cultural values of humans affect the way the natural environment and resources are perceived, used, and managed. Water management principles that heed the local religious context are likely to be more effective than imported, foreign ones. Furthermore, in Muslim countries, developing water management principles that are informed by the teachings of Islam may act as a framework for managing other natural resources. Thus Muslims and non-Muslims need to explore Islam's perspectives on the natural environment in which water resources are recognized as playing a pivotal role. Islamic teachings contain fertile ground for developing water management principles. If applied, perhaps in conjunction with other water management policies in culturally and demographically heterogeneous areas, these principles could find wider acceptance than non-native ones. Such principles would be encouraged by the "penalty and reward" system that is detailed in the Quran and *hadith*.

Rights of the environment

The ultimate objective of life for a Muslim is salvation (Ansari 1994, 397). An Arabic dictionary defines "Islam" as "abiding by obligations and (avoiding) the forbidden without repining." *Salam*, the Arabic root of the word "Islam," means "peace and harmony" (Al Munjid 1994, 347). Ansari (1994, 394), therefore, argues that an "Islamic way of life entails living in peace and harmony" at individual and social as well as ecological levels.

Human-environment interactions exist within dynamic cultural, spatial, and temporal contexts. Given this, it is critical that water management strategies should incorporate elements of local cultures and religions. There are numerous references to water and related phenomena in the Quran. For example, the word "water" (*ma'*) occurs sixty-three times and "river" or "rivers" fifty-two times (Abdul Baqi 1987). Other words such as "fountains," "springs," "rain," "hail," "clouds," and "wind" occur less frequently. Paradise, which, Muslim believe, is the eternal home of believers and those who do righteous deeds,[1] is often depicted in the Quran as having, among other desirable services and objects, running rivers.[2] Furthermore, perhaps the most quoted verse of the Quran is "*And We created from water every living thing.*"[3] It testifies to the centrality of water to life in the ecosystem as a whole, and as the unifying common medium among all species. Given Islam's recognition of water's pivotal importance, a management instrument that broadens traditional (for example, economic) water management approaches to include non-traditional, cultural and spiritual approaches is more likely to succeed in the Muslim world.

In Islam, human-environment interactions are guided by the notion of the person as a *khalifa*, meaning a viceregent or steward of the earth. The philosopher of religion Ali Shariati (d. 1977) argued that the spiritual as well as the material dimensions of humans are both "directed toward the singular human purpose of *khalifa* (viceregency)" (Sonn 1995). Khalid (1996, 20) states that although "we (humans) are equal partners with everything else in the natural world we have added responsibilities. We are decidedly not its lords and masters" but its friends and guardians. One interpretation of *khalifa* is given by Ibn Katheer (1993, 1:75–76). He argues that the *khalifa* should be an adult Muslim male who is just, religiously learned (*mujtahid*), and knowledgeable in warfare. He ought to establish the thresholds (*hudud*) of human conduct as mandated by God, as well as justice and peace among the people. He ought to stand by the oppressed and forbid indecency and despoiling (*fawhish*). Some of the skills of a *khalifa* that were essential fourteen hundred years ago, when

Muslims were under constant threat of attack, are less relevant today – such as knowledge of warfare.

It is impermissible in Islam to abuse one's rights as *khalifa*, because the notion of acting in "good faith" underpins Islamic law. The planet was inherited by all humankind and "all its posterity from generation to generation.... Each generation is only the trustee. No one generation has the right to pollute the planet or consume its natural resources in a manner that leaves for posterity only a polluted planet or one seriously denuded of its resources" (Weeramantry 1988, 61). In other contexts, the concept of *khalifa* refers to the fact that waves of humanity will continuously succeed each other and inherit planet earth.

The Quran enjoins believers to "*Make not mischief on the earth*"[4] and declares that "*Mischief has appeared on land and sea because of (the meed) that the hands of men have earned, that (God) may give them a taste of some of their deeds: in order that they may turn back (from evil).*"[5] When human-produced "mischief" a rough translation (Yusuf Ali 1977) of the Arabic word *fassad* – spoils the natural order, God penalizes people with the same type of affliction that they have inflicted on His creation. The other meanings of *fassad* include taking something unjustifiably and unfairly (Al Munjid 1994) or spoiling or degrading (natural) resources. Tabatabai (1973, 196) views *fassad* as "Anything that spoils the proper functioning of current (natural) regulations of the terrestrial world regardless of whether it was based on the choice of certain people or not.... *Fassad* creates imbalance in the pleasant living of humans." The verses that succeed the passage on *fassad* refer to earth and wind, and to rewards from "*God's bounty*" for those "*who believe (in God) and work righteous deeds.*"[6] The notion of *fassad* is not associated with any specific time and place, and is thus universal and everlasting in scope. *Fassad* is mentioned in the context of "land and sea."[7] It is, however, reasonable to assume that this notion also encompasses all other components of the ecosystem because the Quran states that to God, the creator of everything,[8] belong the heavens and the earth and whatever is between them and what is beneath the ground.[9] Islamic teachings, including the Quran, therefore, command Muslims to avoid and prevent *fassad*, which encompasses undue exploitation or degradation of environmental resources, including water. This perspective is especially revealing in light of the Islamic belief that the natural world is subservient to the human world. Humans are consequently permitted to use and transform the natural environment, with which they are entrusted, to serve their survival needs. For example, God states that humans may use His (good) resources for their sustenance on the condition that they "*commit no excess* (la tatghou) *therein, lest My wrath should justly descend on you.*"[10]

God's "green light" to use water and other resources is conditional on humans' wise and sparing use of it. They ought to employ it to sustain their biological needs. Current users of water and other environmental resources must avoid irreversible damage so that the resources can serve humanity's current and future needs. Muslims are, therefore, permitted to control and manage nature but not to cruelly conquer God's creation.

Being mindful of the needs of current and future generations is an important aspect of piety in Islam. In the words of the *hadith*, "Act in your life as though you are living forever and act for the Hereafter as if you are dying tomorrow" (quoted in Izzi Deen 1990, 194). The *hadith* asks people, in effect, to work for and think of future generations as if they were alive and using these very resources. Just as one would not undermine one's own future, a person ought not rob future generations of their needs.

Muslims are enjoined to "*Violate not the sanctity of the symbols of God*"[11] and to fulfill all of their obligations to Him.[12] In many verses, water and the rest of creation are described as "signs."[13] Different verses in the *Quran* state that these signs are for people who think, hear, see, and have sense, and are intended for the people to give thanks to the Giver. Therefore, one should naturally avoid violating or undermining these divine signs.

Although people are entrusted with caring for the natural world, God states in the Quran that many violate the admittedly heavy burden of trust. In light of this, Islamic teachings state that if one generation of people is "cheated" by preceding ones, it must not cheat succeeding generations. The Prophet said: "Pay the deposit to him who deposited it with you, and do not betray him who betrayed you."[14] A Muslim is instructed to correct environmental failures by abstaining from behaviours that waste or pollute water.

Muslims who engage in *fassad* are effectively sinners. Their environmentally disrupting conduct amounts to breaking "*God's covenant after it is ratified.*"[15] A covenant was "entered into with 'Father Abraham' that in return for God's favours the seed of Abraham would serve God faithfully." At another level, a "similar covenant is metaphorically entered into by every creature of God: for God's loving care, we at least owe Him the fullest gratitude and willing obedience" (Yusuf Ali 1977, n. 45). Therefore, by knowingly violating the teachings of God, one is in effect resisting His grace and sustenance for which one is penalized by, among other things, God withholding his bounty from that person.

The Islamic perspective on the natural environment is holistic. Everything is seen as important, and as interdependent on everything else. God has "*sent down rain from the heavens; and brought forth therewith fruits for your sustenance.*"[16] All environmental media have rights, including a right to water. The Quran, for example, states that "*There is not an animal (that lives) on the earth, nor a being that flies on its wings, but (forms part of) communities like you.*"[17] It also mentions that "*vegetation of* all *kinds*"[18] and of "*various colours*"[19] are nourished by rainwater that God sends down. Water is made available by God in "order that *all* life receives its support according to its needs" (Yusuf Ali 1977, n. 3107 – emphasis added), including humans, animals, and plants.[20] This points to, among other things, the rights of non-human species to sufficient water that is of "good" quality because the water has to be suitable for irrigation and drinking.

Rewards and penalties of Islamic water management

God rewards Muslims who help animals and penalizes those who hurt them (Li Ibn Kadamah 1992; Wescoat 1995). Muslims believe that good deeds annul bad ones[21] and that bad deeds annul good ones. The degree of rewards or penalties for deeds depends upon one's intentions.[22] The Prophet said that "He who amongst you sees something abominable should modify it with the help of his hand; and if he has not strength enough to do it, then he should do it with his tongue; and if he has not strength enough to do it, (even) then he should (abhor it) from his heart."[23]

Similarly, a key directive to and mission for every Muslim is captured in the following Quranic verse, which is repeated in many prayers of Muslims: God "*forbids all shameful deeds, and injustice and rebellion*"[24] (*fahesha*, *munkar*, and *baghi*) against His "law or our own conscience" (Yusuf Ali 1977, n. 2127). "Injustice" can be also understood to include wickedness. Therefore, pollution and wastefulness of natural resources are prohibited because they are unjust in the way they that jeopardize current and future generations' ability to meet their own needs.

Water resources are promised to Muslims who piously abide by the commandments given to them by the Owner of the heavens and earth. Those who follow the straight path as charted by God's message will not "*fall into misery* (shaqa),"[25] nor "*shall they grieve.*"[26] The Quran defines the absence of "misery" (*shaqa*) as people having enough provisions "*for thee not to go hungry nor to go naked. Nor to suffer from thirst, nor from sun's heat.*"[27] The notion of *shaqa* refers to penalties in this life (not in the hereafter), which in turn ought to give Muslims a greater incentive to avoid environmental wickedness. It ought to galvanize Muslims to follow the teachings of their faith in terms of use and management of water resources.

God will provide for pious, practising Muslims abundant water[28] to "test them" in their sustenance and resources[29] (Tabatabai 1974, 20:46). God states that if only humans world have faith and fear Him, "*We should indeed have opened out* (fatahna) *to them all kinds of blessings from heaven and earth*".[30] God also reminds Muslims that it is He who is in control of rain and the one who sends it down.[31] In another chapter,[32] God asks the people the following rhetorical question: Who will bring you flowing surface water if your current supplies become scarce (*na-th'ubet*)? (Tabatabai 1974, 19:365.) Numerous verses and *hadiths* remind Muslims that the resources that they consume daily are ultimately controlled by their Creator. This is reflected in the way that most Muslims commonly use phrases such as "God willing." God's will is a necessary prerequisite for humans and other species to have sufficient water and other resources. Short of this, the "natural" renewability of water is thrown into question. God's will can be humbly appealed to by applying His teachings and message.

The reward and penalty system is designed to induce far more good deeds than bad ones. For example, a "bad deed" counts as one "against" a person who is rewarded ten times to seven hundred times for each righteous deed.[33] Unbelievers are described as having profitlessly "*bartered guidance for error*" and thus having "*lost true direction*" (*huda*).[34] Consequently, only by "living" or applying the teachings of Islam, including its environmental ethic, can people expect replenishment of their diminishing water supplies. This perhaps explains why, when struck by a natural (or human-induced) calamity, many Muslims commonly attribute it to impiety at the individual or societal levels.

The notion of sustenance (*rizq*) occurs frequently in the Quran. It "refers to all that is necessary to sustain and develop life in all its phases, spiritual and mental, as well as physical." (Yusuf Ali 1977, n. 2105).[35] God is believed to be the source of all our sustenance (ibid., n. 5579). Muslims are commanded to reject all rival gods who, according to Yusuf Ali, include idols, poetry, art, science, and pride in wealth (ibid., n. 41). A Muslim should not overvalue the material nor the technological dimensions ("gods") of our modern life because they would distract one from glorifying and worshipping God.

Muslims believe that the faithful who fear God (*itaqu*) follow His guidance, avoid personal temptations (*al-hawa*), do righteous deeds, and avoid evil ones will be rewarded by Him. Good deeds must be within the socio-economic and physical capacity of a Muslim to perform[36] and must be performed on a regular basis.[37] Thus, faithful Muslims will not fall into misery or grief, nor fear for their future. They will be sustained from unexpected sources[38] and admitted to gardens with flowing rivers.[39] Muslims who were once misguided or violated the signs and teachings of God may elect to repent in a genuine way by abiding by the divine instructions. Those who are genuine in their religious belief (*akhlasu*) will be granted rewards of "immense value."[40]

In the preceding section, it was shown that, according to Islam, God, the owner of the natural world, is also its supreme manager who entrusted humans with its stewardship. God will unlock water and other resources for those who abide by His revelations to the Prophet Muhammad. Generally speaking, God rewards the faithful in spiritual or physical ways, and the rewards may take place in this life or in the afterlife. In this life, the rewards include a worry-free existence, and a greater level of water and other resources for sustenance.

Islamic water management institutions

Islam's overall environmental message is one of balance: people should avoid excessive accumulation of material wealth and pride in worldly accomplishments because these sidetrack believers to irreligious temptations, thus disrupting the flow of sustenance. But Islam recognizes the fallibility of humans and their weakness in the face of temptations. It was for this reason that the institution of the *hisba*, the office of public inspection, was created. Throughout much of Islam's history, the *hisba* encompassed both moral issues as well as those touching more widely on everyday life. Today, the moral aspect of the *hisba* no longer exists, except in a few countries such as Saudi Arabia, Iran, and Sudan.

The ethical underpinning of the *hisba* is the *Quran's* instruction about "*enjoining what is right, and forbidding what is wrong*,"[41] and the *sharia* principle of "no injury." The officer in charge of the *hisba* is called the *muhtasib*, and his duty, among other things, is to ensure the proper conduct of people in their public activities, including those involving resources and non-human species. For example, a *muhtasib* is expected to prevent the abuse of animals, protect and manage public land reserves, and regulate water uses (Hamed 1993, 155). According to the great jurist Ibn Taymiah, the most important qualifications of the *muhtasib* are expertise in the subject matter, kindness, and patience. Throughout the Muslim world, the *hisba* should be resurrected and entrusted with the implementation of fair and just water management practices.

Conclusion

The teachings of Islam that advocate wise use of water resources to meet humans' need to sustain themselves can be summarized in the notion of demand management. People, according to Islam, may control nature and consume its resources, but may not cruelly conquer it in such a way as to irreversibly degrade God's creation. Given that a water management strategy that incorporates elements of the "cultural landscape is likely to have a strong impact on the interior landscape" (Orr 1996, 228), policy-makers can tap into Muslims' religiosity and desire for salvation to design and implement an Islamically inspired water management strategy. For Muslims, salvation can be achieved only through applying Islam's teachings and *sharia*, which are clearly water-friendly.

Principles of Islamic water management may be used alone or, as was done in Jordan in the early 1990s, in combination with non-religious slogans on various posters in an effort to induce Jordanians to conserve the kingdom's scarce water resources. Likewise, effective Islamically grounded water policy can be drafted to reflect alternative, non-traditional world views and value systems. Furthermore, sustainable management of water resources in Islamic countries is more likely to be realized if the management instruments incorporate a host of alternative incentives such as religious, spiritual, and resource-based rewards. Culturally sensitive demand management strategies require a deliberate effort of water education about the positive link between Islam and water conservation.

It would be spiritually rewarding at the individual level, and socially and environmentally beneficial at the community or country levels, to educate students in a way that is consistent with their culture and belief system. Hence, Islamic water management principles, when sufficiently developed, should be moved from the academic or religious level to the popular level. The totality of Islamic water management principles, including their educational dimension, ought to change the way Muslims live their lives. God "tests" Muslims by the manner in which they use water (and other) resources. The test is about whether they are "living" their religion by following its principles of conserving water and protecting its quality. Those who do will be rewarded by God with His blessings as well as with increased resources. Disobedient Muslims have opportunities to repent and mend their ways, or will be penalized in this life and the afterlife.

The world cannot be partitioned into "inside" and "outside" spheres, where the natural environment and water resources are the "outside." Humans are embedded in nature and should act as its stewards not its conquerors. Many states in the Muslim world are experiencing serious threats to their water resources: some suffer from drought, and others from floods, poor water quality, and so on. If these threats are not attended to within a culturally meaningful framework, they may spiral into social tensions and, potentially, into violent conflict. The Islamic rules for human-environment relations and the rewards and penalties attached to them are consistent with the very definition of the word "environment," which suggests the active encompassing of the natural, human, and cultural spheres, and some level of reciprocity. In other words, as the Quran and *hadith* teach Muslims, the environment is not a static phenomenon that can be impacted without consequences.

Notes

1. 4:57.
2. For example, 4:73, 5:119, 47:12.
3. 21:30.
4. 2:11.
5. 30:41.
6. 30:42–46.
7. 30:41.
8. 25:2.
9. 20:6; see also 30:26.
10. 20:81.
11. 5:3.
12. 5:1.
13. See, e.g., 16:65, 41:39, for symbols of "the presence and might" of the Creator: 20–27 has a listing of various nature-based signs of God.
14. Abu Dawood 3528.
15. 2:27.
16. 2:22.
17. 6:38.
18. 6:99 (emphasis added).
19. 35:27.
20. 25:49, 26:155, 13:4.
21. 11:114.
22. Al-Bukhari 1.1, 51.
23. Muslim 79.
24. 16:90.
25. 20:123.
26. 2:38.
27. 2:118–19.
28. 7:96, 72:15.
29. 7:15.
30. 7:96.
31. 50:9.
32. 67:31.
33. Al-Bukhari 1.40.
34. 2:16.
35. See, e.g., 2:21, 16:73, 67:21.
36. Al-Bukhari 1.41.
37. 18:46; Al-Bukhari 1.41.
38. 65:3.
39. 2:38, 4:57, 47:12, 65:3.
40. 4:146.
41. 3:104.

REFERENCES

Abdul Baqi, M. F. (1987), *Al Mu'jam al Mufahras li Alfath al Koran al Kareem* [The dictionary of the phrases of the Glorious Quran], Dar al Hadeeth, Cairo.

Al Munjid (1994), *Qamous Al Munjid* (Retrieving dictionary) (34th ed.), Dar el Machreq, Beirut.

Ansari, M. I. (1994), "Islamic Perspectives on Sustainable Development," *American Journal of Islamic Social Science* 11 (3), pp. 394–402.

Hamed, Safei el-Deen (1993), "Seeing the Environment through Islamic Eyes: Application of *Shariah* to Natural Resources Planning and Management," *Journal of Agricultural and Environmental Ethics* 6 (2), pp. 145–64.

Ibn Katheer (1993), *Tafsir al Koran al Ala'theem lil Imam al Hafith Abi al Fida Ismail Ibn Katheer* [Interpretation of the Glorious Quran], Dar al Ma'rifa, Beirut.

Izzi Deen, Mawil (1990), "Environmental Islamic law, Ethics, and Society," in J. R. Engel and J. G. Engel (eds.), *Ethics of Environment and Development: Global Challenge, International Response*, Bellhaven Press, London.

Khalid, F. (1996), "Guardians of the Natural Order," *Our Planet* 8 (2), pp. 18–25.

Li Ibn Kadamah (Abdullah bin Ahmad bin Mohamad bin Kadamah) (1992), *Al Mughnee* [The enricher], Hajr Publishing, Cairo.

Orr, D. W. (1996), "Ecological Literacy," in M. Alan Cahn and R. O'Brien (eds.), *Thinking about the Environment*, M. E. Sharpe, Armonk, N.Y.

Sonn, Tamara (1995), "Tawhid," in *Oxford Encyclopaedia of the Modern Islamic World*, Oxford University Press, Oxford.

Tabatabai, M. H. (1973), *Al-Mizan fi Tafsir al-Kor'an* [The tempered interpretation of the Koran], vol. 16, Al Alami Library, Beirut.

——— (1974), *Al-Mizan fi Tafsir al-Kor'an* [The tempered interpretation of the Koran], vols. 19–20, Al Alami Library, Beirut.

Weeramantry, C. G. (1988), *Islamic Jurisprudence: An International Perspective*, St. Martin's Press, New York.

Wescoat, J. L., Jr. (1995), "The Right of Thirst for Animals in Islamic Law: A Comparative Approach," *Environment and Planning: D – Society and Space* 13 (2), pp. 637–54.

Vidart, D. (1978), "Environmental Education: Theory and Practice," *Prospects* 8 (4), pp. 466–79.

Yusuf Ali, A. (1977), *The Holy Qur'an: Text, Translation, and Commentary*, American Trust Publications for The Muslim Student Association of the United States and Canada, Plainfield, Ind.

4

Water conservation through public awareness based on Islamic teachings in the Eastern Mediterranean Region

Sadok Atallah, M. Z. Ali Khan, and Mazen Malkawi

Water conservation is a complex interconnecting system with a variety of aspects – from consumer education to advanced technological equipment. All these aspects must be considered in relation to their economic, social, religious, political, legal, and aesthetic contexts (Khan and Abdul Razzak 1986; Abdul Razzak and Khan 1990). Water conservation must be seen as a basic component of integrated water resources management, and public awareness and education are basic tools needed to guarantee the participation and involvement of the public in water conservation (WMO 1992; UN 1993a, b). This is of particular importance in the World Health Organization (WHO) Eastern Mediterranean Region (EMR) which comprises twenty-three countries,[1] most of them situated in arid or semi-arid zones with low annual rainfall, and with a combined population of about 436 million, most of them Muslims. Hence, the aim of this chapter is to outline the importance in the EMR of using the Islamic administration, education system, and teachings in water conservation, especially in increasing and improving the participation and awareness of the public in conserving water.

Water conservation from an Islamic perspective

It has been shown over the last 10 years that campaigning for the conservation of the environment within the Islamic faith is productive, and specifically that using the Islamic education system to address the public of the EMR on important issues such as water conservation has a beneficial effect in raising public awareness. Hamdan et al. (1997, 241) concluded that "there is a desperate need for Islamic environmentalism in our finite world." Raising awareness using Islamic concepts of water conservation is feasible for the following reasons:

- Islam has a strong influence in the EMR;
- Water conservation and protection are stressed in Islamic teachings; and
- Islamic communication channels are very effective in raising public awareness.

The influence of Islam

The use of Islamic concepts to promote human well-being in all aspects of life is common throughout the region. For example, leading physicians, scientists, jurists, and religious scholars from the twenty-three countries of the EMR, meeting in Amman in 1996, declared the importance of Islamic behaviour for promoting good health (WHO 1996a, b, c). This important meeting realized the importance of looking at health as one element of life that cannot be achieved except in conjunction with other basic elements such as freedom, security, justice, water, and food. The meeting also focused on the effect of lifestyle and individual behaviour on health. Islam promotes behaviours that protect health and discourages habits that have a detrimental effect on it. The declaration identified sixty lifestyles where Islamic teachings offer guidance on healthy and harmful behaviours. Water conservation and protection were among the areas of concern.

Water conservation and protection in Islamic teachings

In Islam, the relationship between humans and water is part of daily social existence, which is based on the Muslim belief that everything on earth worships the same God. This worship is not merely ritual practice, because rituals are simply the symbolic human manifestation of submission to God. Rather, worship consists of actions that can be performed by all creatures that share the planet with the human race. Moreover, humans are responsible for the welfare and sustenance of the other citizens of this global environment. Water is the most precious and valuable resource of the physical environment for all living things. The link between life and water is explicitly stated in several verses of the Holy Quran, for example, "*We made from water everything*";[2] "*And Allah sends down rain from the skies, and gives therewith life to the earth after its death.*"[3]

Islam places strong emphasis on the achievement of perfect harmony between spiritual and physical purification. Physical purification cannot be achieved except by ablution and bathing (*ghusl*), which both require clean water. Therefore, purity and cleanliness of water receive a great deal of attention in both the Holy Quran and the *sunnah*, and Muslims are urged not to pollute water. "The Messenger of Allah forbade to urinate in stagnant water";[4] "Let no one of you bathe in stagnant water to remove the state of ceremonial impurity";[5] and "Guard against the three practices which invite people's curses: evacuating one's bowels near water sources, by the roadside and in the shade."[6]

Conservation is a fixed concept in Islamic teaching. It is a way of living that should be implemented through the Muslim's whole life: not as an ad hoc solution to shortages, nor in occasional situations (Madani 1989), but at all times, both good and bad. Islamic teachings tend to emphasize adherence to balance and the just satisfaction of individual and group desires and needs. Such teachings are based on various texts of the Holy Quran: "*O children of Adam! Wear your beautiful apparel at every time and place of prayer: eat and drink: but waste not by excess for Allah loveth not wasters*"[7] or "*Verily spendthrifts are brothers of the Evil Ones; and the Evil One is to his Lord (Himself) ungrateful.*"[8] This is valid for all natural resources. However, Islam gives special attention to water conservation. According to the tradition of the Prophet of Islam, a Muslim is ordered to be economical with water even if he is taking his water from a fast-flowing river. "Allah's Apostle (peace be upon him) happened to pass by Sa'd as he was performing ablution. Whereupon he said: Sa'd what is this extravagance? He said: Can there be any idea of extravagance in ablution? Whereupon he (the Prophet) said: Yes, even if you are by the side of a flowing river."[9]

Other Islamic concepts and principles, such as human appointment and viceregency, co-operation and public participation, public consultation, and the relation between the public and the governing bodies are well documented in the Islamic teachings, and are useful tools for raising awareness and involving the public in water resource management and conservation. Water conservation is not the sole preserve of water agencies; everyone must participate in fulfilment of the Quranic injunction, "*Help ye one another in righteousness and piety. But help ye not one another in sin and rancour.*"[10] The whole planet has been placed under human responsibility to be cared for and not misused. Furthermore, Muslims believe that God created the human race for a great reason, that it might act as His viceroy upon the earth. Our mastery of the earth is for its betterment and development and not for evil or misuse. "*Behold thy Lord said to the angels: 'I will create a viceroy on earth.' They said 'Wilt Thou place therein one who will make mischief therein and shed blood?*

Whilst we celebrate Thy praises and glorify Thy holy (name)?' He said: 'I know that ye know not'."[11]

"Enjoining of good and forbidding evil" is an obligatory action (*wajeb*) that should be performed by all Muslims. It is an important tool not only for raising awareness, but also for urging good action and involvement and prohibiting bad and harmful behaviour, for the Prophet (pbuh) "*commands ... what is just and forbids them what is evil.*"[12] The principle of "Neither harm nor harming"[13] is another basis for declaring an official Islamic position toward water conservation issues. It can be interpreted from this *hadith* that all harmful acts are forbidden (*haraam*).

Typical Islamic behaviour and action are guided by the *hadith*, "I heard the messenger of Allah as saying: He who amongst you sees something abominable should modify it with the help of his hand; and if he has not strength enough to do it, then he should do it with his tongue; and if he has not strength enough to do it, (even) then he should (abhor it) from his heart and that is the least faith."[14]

Public awareness through Islamic communication channels

Muslims believe that Islam gives meaning to individual human activities and to society as a whole. One's belief in Islam may be viewed as a reflection of society: or the state of society may be regarded as depending upon the beliefs and actions of the individuals within it. Like other religious groups, Muslims believe that the morality of individuals within a society is its basic building block. Morality overrides all material benefits that a Muslim stands to gain, and thus represents the necessary basis for that society (Hamdan et al. 1997).

The Islamic education system offers several fora for delivering Islamic teachings. The mosque is the best forum for addressing the general public at all levels on matters covering all the issues of daily life. At a minimum, the Friday prayer represents a weekly opportunity to address the public. However, in most Muslim countries, there are daily gatherings where imams can address people on issues that are felt to be important. In a survey conducted in Amman, 64 per cent of respondents thought that imams had an important role in environmental education and public awareness – although only 34 per cent indicated that imams were already filling that role (Al-Sodi 1993).

In Islam, everyone is responsible for education, from the family level up to the whole society. Although this applies to all aspects of life, the mufti of Jordan delivered a specific *fatwa* that environmental education is *wajeb*, or an obligation: under such a *fatwa*, all Muslims are responsible for participation in environmental education. This *fatwa* is based on a basic Islamic rule that "whatever is mandatory for completing *wajeb* is *wajeb*" (Al-Tamimi 1991). Therefore, Islam provides a dynamic forum that is capable of reaching the entire Muslim population – in the house, street, school, and mosque.

This universal responsibility for education provides ideal tools and fora for reaching the public in Muslim countries. Unfortunately, these have not been used efficiently. In the few countries where religion has been used to support public awareness campaigns, this has been limited to the use of some texts from the Holy Quran and *sunnah* in posters and articles in the newspapers. But water-conservation must involve all people and requires behavioural changes. It involves sacrifices and social and financial costs, which necessitate the full co-operation and integration of efforts of all stake-holders. Hence, isolated activities will not achieve tangible results. What is needed is water resources management and conservation strategies and plans that incorporate Islamic concepts and tools in their public awareness activities. This chapter attempts to provide a mechanism and some guidelines to help interested agencies adopt and implement effective public awareness programs, strategies, and activities based on Islamic concepts.

Water conservation and public awareness in the EMR

At the governmental level in most EMR countries, there is a strong belief that water conservation is one of the most reliable and cost-effective solutions to the water shortage problem faced by the region. This belief is clearly reflected in the recommendations made by regional and international meetings of water agencies and international bodies (WHO 1992, 1995, 1997; USAID 1993, World Bank 1995; UNIDO 1997). Unfortunately, it is limited to the producers of water (water agencies and decision makers) and has not been conveyed to the consumers (the public). Lack of public participation and poor awareness on the part of consumers seem to be the main reasons behind this gap.

A comprehensive search was done for literature on water conservation public awareness activities in the EMR but few references were retrieved. This highlights two basic problems: first, the lack of such activities; and, second, the poor information exchange and accessibility in this important sector. A further major problem is that most of conservation activities have so far been targeted at domestic users, with very little focus on agriculture and industry.

Regional and intercountry activities

The WHO Regional Centre for Environmental Health Activities (CEHA) is very active in promoting integrated water resources management as an optimal approach for improving access to safe water supplies in the

EMR. Water conservation is being addressed as an integral part of water resources management. Two intercountry meetings and several national meetings been convened, and several special studies have been carried out, since 1991. Outputs include a draft water conservation strategy and commitment to mobilize national resources toward conserving water resources. Currently, efforts are focused on the development of a modular water conservation handbook, which will consist of ten modules covering all related aspects including public awareness.

The WHO has had extensive experience in raising public awareness and educating the public in the EMR through integrating Islamic teachings as part of the health education program and materials. For example, the WHO Eastern Mediterranean Regional Office (EMRO) launched a program entitled "The Right Path to Health: Health Education through Religion." In this program, both environmental health and water and sanitation issues were addressed because of their importance in the region.

National activities

Four examples, from Afghanistan, the Gulf Cooperation Council (GCC) countries, Egypt, and Jordan, demonstrate the need for and effectiveness of using Islamic concepts to raise public awareness.

In Afghanistan, the WHO launched the "Health Education and Awareness through Mosques" campaign in late 1997. It was the first of a series of campaigns that will be carried out in Afghani cities to promote good health practices and to raise awareness about water conservation and the importance of safe water, adequate sanitation, and hygiene in disease prevention. In this campaign, training of imams by water and religion experts was an essential component. They were provided with clear messages quoted from the available literature. Upon receiving the proper training, each imam prepared a special Friday speech and delivered it twice at two consecutive Friday prayers (WHO 1997). Initial evaluation showed that the messages were well received. However, further evaluation is expected upon completion of the campaign.

In the GCC countries, a shift in societal values from a development-oriented to a conservation-oriented view of water resources is occurring. It is believed in the GCC countries that conservation of natural resources in general, and water resources in particular, is a principal component of Islamic teachings. It is also believed that the most important and effective way to make the public aware of conservation from an Islamic perspective is through the media and the educational system (Akkad 1990). Islamic messages are being used in the preparation of posters and video clippings for these campaigns. On the occasion of World Water Day 1998, and upon request of ministries of Islamic Affairs, imams were requested to devote their Friday speeches to the theme of Islam and water conservation (Salih 1998). But such occasional public awareness water conservation campaigns need to be integrated through a comprehensive and long-term plan of action that targets behavioural change, otherwise their effect will limited.

The National Community Water Conservation Programme (NCWCP) in Egypt was created to address problems of potable water loss, mainly using conservation activities at the national and local level. The NCWCP implemented massive communication activities in 1993–96. According to Afifi (1996, 7), one of the main lessons was that "the strategy of water conservation communication must be global and interactive, and include all consumers and all the factors concerned, such as religious, political and informal community leaders."

In Jordan, a project to make water of improved quality available in increased quantity on a sustainable basis is in course of implementation. A major part of this project involves public awareness activities. Various education and awareness materials (posters, games, newspaper reports, television programmes, seminars, and so on) have been prepared using Islamic teachings and concepts. Several Friday prayers were devoted to water and conservation issues (Ayesh 1996). In collaboration with the Ministry of *Awqaf* and Islamic Affairs, a pilot project titled "Week of the Mosque" was implemented in early 1998. Imams in all the mosques of Amman Governorate were trained for one week to incorporate issues of daily life, including water conservation, into their Islamic education. They were provided with information about Jordan's water resources and the shortage the country faced, and about the need for public co-operation and participation in water conservation. The imams then started educating the public. It is planned to replicate this activity in the other governorates of Jordan.

Islamic water conservation strategies

Water conservation programs should be carried out by the agency responsible for water resources management. The execution of such plans for the various sectors (such as municipal, agricultural, or industrial) should be co-ordinated by the relevant government body in each sector. Close co-ordination and partnership should be institutionalized between the agencies responsible for water supply, demand management, and education, media, and awareness. Unfortunately, the ministries of Education, *Awqaf*, and Islamic Affairs rarely participate in water conservation programmes in the region, although this is essential for effective awareness activities.

Equally, the public's involvement and its co-operation in designing and implementing conservation measures are essential to the success of water conservation programmes. The public includes consumers, service providers, managers, and planners as well as policy-makers. Raising public awareness using Islamic conservation concepts should always be integrated with the use of other communication tools and channels.

To achieve greater co-operation and involvement, the public must understand the water supply situation, including the cost of delivery, the overall water resources situation, and the need to conserve water resources and to maintain them for future generations. This increased understanding is the first step in any successful public awareness activity. However, the credibility of this information is essential. Because honesty is a core principle in Islam, the public expects the truth from imams and other Islamic sources.

Most water conservation activities require changes of behaviour and attitudes, which is usually a slow process. Therefore, ad hoc and occasional public awareness activities are not effective. Water authorities should plan continuous, long-term activities in close collaboration and co-ordination with ministries of Education and Islamic Affairs.

Some water conservation activities involve costs that must be paid by the public, such as fixing water taps, upgrading irrigation systems, or modifying industrial production lines. These costs of water conservation programmes must be offset by some incentives. In addition to the physical incentives, the spiritual incentives offered by Islam can be of value.

Although conservation, co-operation, and other concepts related to water conservation are well defined in Islam, it would help if these concepts were documented and if the regional and national Islamic legislative institutions issued an official Islamic position (*fatwa*). Such a *fatwa* would likely lead to greater conservation because stating that wasting water is *haraam* carries greater impact than simply discouraging the waste of water.

Water conservation activities and awareness campaigns typically focus solely on domestic users. This is shortsighted, and the focus should be on all water users. Mosques are ideal places for awareness campaigns, since all kinds of people meet there at least weekly. However, imams should be aware of the need to address all sections of the population.

Raising awareness through mosques

Any knowledgeable Muslim can educate others about Islam. Although there is no formal clergy in Islam, imams play a key role in delivering Islamic teachings and educating the public through the mosques. Therefore, imams and mosques should be the focus for public awareness activities on water conservation. Imams should be properly trained and informed: as community leaders, they should never be excluded from water resources planning and management activities.

Imams are more capable of reaching the public than water specialists. Although they are usually well educated about *fiqh*, *sunnah*, and *sharia*, their knowledge of water resources and conservation practices is usually insufficient for them to act as educators on the subject. Therefore, water specialists must train, educate, and inform imams not only about water shortages, water conservation practices, and the need to involve the public but also about audiovisual tools and materials to help them reach the public.

Friday prayers, held in the mosques, are an important weekly occasion in the life of a Muslim. Imams should prepare their Friday speeches in close collaboration with water conservation and communication experts, using reliable facts and figures. A Friday speech on a water-related topic should not be occasional, but should be reasonably frequent to achieve a change in behaviour. It is recommended that such speeches be made more often in summer and during periods where water demand is at its peak.

Raising awareness through formal and informal education

In addition to the mosques, all levels of formal and informal education are essential to raising awareness. The topics of water shortage and conservation can be addressed in the course of teaching subjects such as religion, Arabic, science, and geography. Unfortunately, environmental education is in its early phases in most of the EMR countries. Therefore, the ultimate goal will be to upgrade curricula in the subjects mentioned to include environmental education, which should address all the priority issues including water resources, protection, and conservation. This upgrading of the curricula will take some time. Therefore, whenever changes are being made to textbooks or curricula, water authorities should make use of such chances by making sure that water conservation concepts are included.

In view of the poor formal environmental education, informal education seems to be more feasible in the short run. Seminars, workshops, and lectures should be arranged for students and other groups. It is important to incorporate Islamic concepts in such activities. Reports and articles in newspapers, television shots, posters, and other mass media tools are also very effective in addressing the general public. Usually this is the responsibility of water authorities in collaboration with other environment protection agencies. However, it is important to incorporate Islamic concepts and attitudes in these tools. This of course has to be done in close collaboration with ministries of *Awqaf* and Islamic Affairs.

Conclusions

By the year 2050, almost all the EMR countries will be facing water shortages. Integrated water resources management is the most feasible option to overcome this serious crisis. Water conservation should be an integral part of this option, with a clear focus on public awareness and participation without which the chances of success will be lessened.

Access to information related to water conservation and public awareness activities is lacking in countries of the EMR both because of the limited number of these activities, and because of poor information management and exchange. Such access can be improved by compiling a database to identify all the available literature documenting the experience of the EMR countries in water conservation and public awareness. This database should be made available to water specialists as well as to the public by networking at the regional and national levels.

Although Islam discusses issues such as conservation, co-operation, preventing harm, and water pollution protection, the official Islamic position on these issues in relation to conserving water resources needs to be proclaimed by regional Islamic legislative bodies. This will require close co-operation and co-ordination between water agencies and the legislative bodies.

Islam is influential in the EMR countries, and Islamic behaviour can help to achieve health and well-being. Incorporating Islamic teachings on water conservation into the speeches of imams, education, and the mass media will help raise public awareness of the need to manage water scarcity. However, this must be done in close co-ordination and collaboration with all the stake-holders, and should be integrated into overall water resources management.

Water-conservation activities require behavioural changes. Such changes are usually slow. Therefore, long-term plans of action require long-term projects because occasional, isolated efforts are ineffective. Pilot public awareness projects should be initiated, undergo appropriate modification, and then replicated and maintained on a larger scale.

Notes

1. Afghanistan, Bahrain, Cyprus, Djibouti, Egypt, Iran, Iraq, Jordan, Kuwait, Lebanon, Libya, Morocco, Oman, Pakistan, Palestine, Qatar, Saudi Arabia, Somalia, Syria, Sudan, Tunisia, United Arab Emirates, Yemen.
2. 21:30.
3. 16:65.
4. Muslim 553.
5. Muslim 423, in *Hadith Encyclopedia.*

6. Abu-Dawood 24, in *Hadith Encyclopedia.*
7. 7:31.
8. Al-Israa: 27
9. Al-Termithi 427.
10. 5:2.
11. 2:30.
12. 7:157.
13. Ibn-Majah, in *Hadith Encyclopedia.*
14. Muslim 79.

REFERENCES

Abdul Razzak, M. J. and Khan, M. Z. A. (1990), "Domestic Water Conservation Potential in Saudi Arabia," *Journal of Environmental Management* 14 (2), pp. 167–78.

Afifi, Madiha Moustafa (1996), "Egyptian National Community Water Conservation Programme," in *Environmental Communication Strategy and Planning for NGOs, Ma'ain, Jordan, 27–31 May 1996*, Jordan Environment Society, Amman.

Akkad, A. A. (1990), "Water Conservation in Arabian Gulf Countries," *Journal of the American Water Works Association* 82 (5), pp. 40–50.

Al-Sodi, Abdul Mahdi (1993), *Attitudes of Jordanian Citizens towards Environmental Protection in the Sweileh and Naser Mountains Areas* (in Arabic), Environmental Research and Studies, 3, Jordan Environment Society, Amman.

Al-Tamimi, Izz El Din (1991), *Religion As a Power for Protection of the Environment* (in Arabic), Environmental Research and Studies, 1, Jordan Environment Society, Amman.

Ayesh, Mohammed (1996), "Awareness Project in Water," in *Environmental Communication Strategy and Planning for NGOs, Ma'ain, Jordan, 27–31 May 1996*, Jordan Environment Society, Amman.

Khan, M. Z. A. and Abdul Razzak, M. J. (1986), "Domestic Water Conservation Technology in Arid Regions," *Arabian Journal for Science and Engineering*, 2 (4).

Hamdan, M., Toukan, Ali, Shaniek, M., Abu Zaki, M., Abu Sharar, T., and Saqqar, M. (1997), "Environment and Islamic Education," in *International Conference on Role of Islam in Environmental Conservation and Protection, 22–23 May 1997*, Al-Najah University, Nablus, Palestine.

Madani, Ismail (1989), "Islam and Environment," in *For Environmental Awareness in the Gulf Countries*, Ministry of Information, Manama, Bahrain.

Salih, Abdin (1998), *Qatar IHP Committee Celebrates World Water Day.* Http:// Waterway.org.

Samarrai, Mawil Izzi Dien (1993), *Sharia'a and Environment*, University of Wales, Lampter.

UN (United Nations) (1993a), *Agenda 21, Chapter 18: Protection of the Quality and Supply of Freshwater: Application of Integrated Approaches to the Development, Management and Use of Water Resources*, International Development Research Centre, Ottawa.

——— (1993b), *Agenda 21, Chapter 36: Promoting Education, Public Awareness and Training*, International Development Research Centre, Ottawa.

UNIDO (United Nations Industrial Development Organization) (1997), "The Role of Industry in the Development and Conservation of Water Resources in the Arab Region: Challenges and Prospects," in *Workshop on the Role of Industry in the Development of Rational Use of Water Resources in the Middle East and North Africa, Amman, Jordan, 13–15 May 1996*, UNIDO, Vienna.

USAID (US Agency for International Development) (1993), *Water Resources Action Plan for the Near East*, USAID, Washington, D.C.

WHO (World Health Organization) (1992), *Regional Strategy for Health and Environment*, WHO/EMRO/CEHA, Amman.

WHO (World Health Organization) (1995), *Beirut Declaration on Action for a Healthy Environment.* WHO/EMRO/CEHA, Amman.

——— (1996a), *Health Promotion through Islamic Lifestyles: The Amman Declaration*, The Right Path to Health: Health Education through Religion, 5, WHO/EMRO, Alcxandria.

——— (1996b), *Water and Sanitation in Islam*, The Right Path to Health: Health Education through Religion, 2, WHO/EMRO, Alexandria.

——— (1996c), *Environmental Health in the Islamic Perspective* (in Arabic), The Right Path to Health: Health Education through Religion, 7, WHO/EMRO, Alexandria.

——— (1997), *Centre for Environmental Health Activities (CEHA) News Letter*, no. 23, WHO/EMRO/CEHA, Amman.

WMO (World Meteorological Organization) (1992), *The Dublin Statement. International Conference on Water and Environment: Development Issues of the Twenty-First Century*, 26–31 January 1992, WMO, Geneva.

World Bank (1995), *From Scarcity to Security: Averting a Water Crisis in the Middle East and North Africa.* World Bank, Washington, D.C.

5

Water conservation through community institutions in Pakistan: Mosques and religious schools

S. M. S. Shah, M. A. Baig, A. A. Khan, and H. F. Gabriel

Water is the critical resource for humanity. No life can exist without water. In the Quran, Allah says, "*We made from water every living thing*."[1] Water is essential not only for the survival of human beings but also for animals, plants and other living beings. It is a fixed quantity on this planet, as the Quran says: "*And We send down water from the sky according to (due) measure and We cause it to soak in the soil; and We certainly are able to drain it off (with ease)*."[2] The Quran strongly discourages the waste of any resource, including water: "*But waste not by excess: for Allah loveth not the wasters*."[3] Water wastage in particular is strongly discouraged, as is clear from the following *hadith*. The Prophet Muhammad's (pbuh) wife, Ayeshah, said that "the Prophet used to use a very small quantity [equal to $\frac{2}{3}$ litre] for ablution and a bit more [equal to 2–$3\frac{1}{2}$ litres] for bathing."[4] Ablution is the process of washing that is required of every Muslim before prayers and, even in this important task, the Prophet set an example by conserving water. Thus, water conservation in Muslim countries can be planned on the basis of Islam. Such plans will be effective because they rely on a natural approach to handling water scarcity and they will produce much better results than if plans are based only on government regulations.

This chapter highlights the use of mosques and religious schools to promote water conservation by linking it with Islamic teachings. Because water is used in different ways in different times and places; it will not be possible to use this approach in all cases, but often it will be very effective, as the data from two informal case studies in a small town and a village of Pakistan demonstrate. Of the three main types of water consumption, municipal supply, irrigation, and industrial supply, the first two are vital in any country because they involve human consumption of water and the cultivation of crops. In these two categories of water use, mosques and religious schools can play a crucial role in managing and conserving available water resources through a detailed plan based on established practices and Islamic teachings.

Municipal water supply

In the middle of 1991, a few educated people in a small town of the Dijkot District, Faisalabad, formed a group to solve the local water shortage problem by reducing losses and optimising use of water. The group was not a non-governmental organisation (NGO): as such; rather, it consisted simply of people who met at the mosque quite regularly for their prayers. One of them had the idea of trying to use the mosques and religious schools (*madaris*) to influence people's thinking and behaviour regarding the wastage of municipal and irrigation water supplies. The idea was first to motivate the people in the town of Dijkot for optimal water use and control of wastage, and then to apply the experience in a nearby village to solve the problem of the shortage of irrigation water for crops. The group asked the lead author of this chapter to help as a guide or team leader and work for the cause on the weekends, and as Dijkot is his hometown, he agreed.

It was decided to use a very simple, but systematic approach to the problem. The first step was to conduct a house-to-house survey regarding shortage of water so as to be able to assess the severity of the problem. The survey was conducted in the evening because the personnel were volunteers who had to work during the day. The following simple questions were asked:

- Are you satisfied with the available water supply?
- If not, what could be the reason for the problem, and how do you think it could be remedied?
- What expectations do you have of the local government officials responsible for water supply?

Consumers in a total of 4,113 houses were surveyed, in four groups (table 1).

- The first group, about 30 per cent of the total, lived close to the town's main water supply reservoir. They had no complaints because they were getting the required quantity of water.

Table 1 **Number of houses grouped by distance from water supply tank, and numbers in each group experiencing water shortages before and after implementation of the action plan**

	Total	With water shortages	
		Before (per cent)	After (per cent)
First group	1,234	0	
Second group	1,028	0[a]	
Third group	823	412 (50)	164 (20)
Fourth group	1,028	771 (75)	433 (42)
Total (third and fourth groups)	1,851	1,183 (64)	597 (32)
Total	4,113		

[a] Assumed.

- The second group, about 25 per cent of the total, lived a short distance from the reservoir. They mentioned various minor problems of shortage but were reluctant to explain how they had overcome these.
- The third group, 20 per cent of the total, lived farther from the reservoir. The people in this group complained because almost 50 per cent of the houses in this group were experiencing water shortages that the consumers blamed on government officials.
- The final group, 25 per cent of the total, lived on the periphery of the water supply network – in other words they were tail-end users. They were experiencing a serious water shortage, such that 20–25 per cent of the houses were hardly receiving any municipal water at all. From time to time the consumers complained and protested to the local government officials.

Detailed data analysis and interpretation through discussions gave the following information.

- The first group of consumers had no problems since they lived near to the water source; rather they were wasting water by unnecessarily leaving taps running.
- The second group was not facing serious problems because they were using pumps connected to the supply pipes so as to draw water illegally to overcome any deficiency. This, of course, was why they were reluctant to talk about how they overcame shortages.
- The consumers in the third group were also using illegal pumps to draw water, but only 50 per cent of the houses could meet their requirements.
- The fourth group was also trying to use illegal pumps but even then only 20–25 per cent were successful. Almost 75 per cent of the houses were experiencing serious water shortages.

Following detailed discussions, the research group decided to mount an information and awareness campaign targeted at the first two groups and part of the third group. Because the problems differed in each group, different themes had to be stressed in each group.

- For the first group, the campaign would be directed against wastage of water;
- For the second group, the campaign would be mainly directed against illegal water pumps and supply connections, and only partly against water wastage; and
- For the targeted part of third group, the campaign would deal exclusively with illegal pumps.

Influential imams of selected mosques in the relevant localities were asked to discuss the problem in their *khutba* (Friday speeches). A few of them would not agree to co-operate, but most were willing. Relevant literature, based on moral values and religious points of view, was distributed so as give the imams information to supplement what they already had, so that their Friday sermons would be more effective. Volunteers from religious schools were asked to prepare handwritten posters, highlighting religious and moral condemnation of water wastage and the use of illegal pumps and connections. The intention was to point out to the culprits that taking another person's share of water is sinful.

Almost three months were spent in collecting literature and preparing material, mainly because relevant literature was unavailable or inaccessible. Actually implementing the action plan was the most difficult job. It took almost six months because religious students could only work as helpers work once a week, for no more than half a day. The imams suggested that it would not be effective if every *khutba* were on this topic, so speeches of thirty or forty minutes were delivered twice a month. The total involvement of the mosque imams and students of religious schools, in terms of hours worked, was not calculated, but their assistance was provided over a period of nine or ten months.

Records of *khutba* are not available as they were not in written form, and unfortunately, they could not be tape-recorded. The imams were given only the main points and details of presentation were totally in their hands. Thus, useful material that could be used for future projects or studies cannot be documented.

Two months after the plan was implemented, a second simple survey was conducted for assessment purposes. This time, only the third and fourth groups were surveyed because they were the ones suffering from water shortages. In terms of numbers of complaints, the improvement after implementation was a reduction from 50 per cent complaining to almost 20 per cent in the third group and from 75 per cent to 42 per cent in the fourth group. The overall rate of complaints among all houses experiencing water shortages in both groups decreased from 64 per cent to about 32 per cent. Thus about half of houses that had been facing water shortages before were no longer facing problems (table 1).

The campaign highlighted the ineffectiveness of official actions to combat the water shortage. Government officials had sent a general warning to all communities regarding the wastage of water, irrespective of the real culprits. This warning was a serious mistake because it did not make the culprits realize their guilt – they considered it as a general warning and not specific to them. At the same time, the general warning irritated the main sufferers – they were not doing anything wrong but were still being admonished. Further, the government was charging a flat rate irrespective of the quantity of water being used, and penalties for illegal pumps and connections were too trivial to act as deterrents. Complaints or protests from the tail-end users, the main sufferers, were generally ignored by officials who could not take serious action against users of illegal pumps and connections because the procedure for even minor penalties was too complex, and the culprits sometimes used political pressure to escape prosecution.

Just as important as the success of the campaign were the various problems of implementing measures against the water shortage that were identified:

- Government rules and regulations not based on realities on the ground;
- Illiteracy of the population;
- Lack of awareness of the problem;
- Lack of clear understanding by religious leaders of the issue of water conservation in which they could play a crucial role to improve the situation;
- Lack of interest by government officials in investigating the problem in depth;
- Unavailability of dedicated full-time volunteers for awareness campaigns; and
- Lack of professional skills of those who did volunteer.

Irrigation water

In a nearby village, a case study was conducted on similar lines. The problem involved a side canal leading water from a main canal by gravity flow; the farmers were supposed to be supplied with water from the side canal on a time-sharing basis based on the size of their land holdings. A survey similar to the one conducted in the town of Dijkot found that the problems were more or less the same. The tail-end users of the side canal were desperate for water whereas landowners at the middle and head of the canal were enjoying sufficient water, and were even stealing water in various ways in case of any deficiency. The weaknesses of the government rules and regulations were also similar. For example, flat rates were charged, based on time sharing related to the size of the land holding, whether or not the farmer in fact received his full share of water. In addition, the farmer had to take his share whether he needed it or not – if it was not needed, then it was wasted. Likewise, the penalties for stealing and illegally diverting water from the canal were trivial.

A similar action plan was developed, except that in this case, the village headman (*lumberdar*) and heads of influential families were involved instead of the imams.

The difficulties and limitations in implementing the action plan in the village were more or less the same as in the town. To assess success, a similar technique was employed, which revealed a lower success rate than in the town. Even so, there were about 26 per cent fewer complaints of inadequate water after the implementation of the plan than before it.

Conclusions

Because of the limitations of these case studies in terms of time and resources, further, more methodical and scientific studies should be conducted to clearly establish the link between reduction of water use and public awareness programs based, in part, on religious values. Despite the limitations, the following conclusions can be drawn from the case studies.

- Mosques and religious schools, using religious points of view regarding the wastage of water, can play a useful role in controlling water wastage: the reduction of water shortage complaints found in these case studies suggests that savings of water can be significant.
- Government rules and regulations alone are often ineffective in changing people's water management behaviour.
- For sustainability, long-term policies are needed.
- For best results, NGOs and government agencies should work as partners.
- NGO members who work with imams and religious schools on public awareness programs in water conservation require professional skills.

Although the case study was not formal and time was limited, a lot of experience was gained in using a religious approach for future planning in areas with similar water shortage problems. In any Muslim country, public awareness programs based upon Islamic teachings about conservation should include the following components to achieve long-term, sustainable benefits:

- Introducing courses on water management and conservation, based on an Islamic viewpoint, and specifically the Quran and *hadith*, in the syllabuses of religious institutions.
- Allocating financial resources in the religious studies departments of universities as well as in religious institutions to support research at the postgraduate level on the Islamic viewpoint on water management and conservation;
- Conducting short courses and workshops to educate government officials on the Islamic viewpoint regarding water conservation, using the research work done in the religious studies departments of universities and in religious institutions;
- Training of the students of religious institutions in the use of solid religious arguments to influence public thinking and behaviour on water wastage and conservation issues;
- Developing NGOs that involve local religious leaders and students, and supporting them with government funds to ensure continuity and sustainability of their work; and
- Basing government rules and regulations on realities on the ground and on the suggestions of NGOs working in the local communities. Water conservation awareness campaigns must be launched by NGOs in their local communities.
- Model studies should be started in several different cities, if possible in different Muslim countries. In this way, experience from the various studies could be used to improve the plans. Preferably, this process should start in Muslim countries where literacy is high.

Notes

1. 21:30.
2. 23:18.
3. 6:141.
4. Al-Bukhari 1.200.

6

Water demand management in Saudi Arabia

Walid A. Abderrahman

Abdullah bin 'Umar said, "I heard Allah's Apostle saying, 'All of you are guardians and responsible for your wards and the things under your care. The imam (that is, ruler) is the guardian of his subjects and is responsible for them and a man is the guardian of his family and is responsible for them. A woman is the guardian of her husband's house and is responsible for it. A servant is the guardian of his master's belongings and is responsible for them.' I thought that he also said, 'A man is the guardian of his father's property and is responsible for it. All of you are guardians and responsible for your wards and the things under your care.'"[1] This *hadith* indicates clearly the responsibility of governments to secure basic needs such as water for the people.

Understanding its duties, the government of Saudi Arabia, which follows the principles of Islamic law, or *sharia*, in all aspects of life, founded specialized water agencies for production, distribution, and treatment of water in the kingdom in the post–World War II decades. The Ministry of Agriculture and Water (MAW) was established in 1953, and was assigned the responsibility for water production to satisfy the required water demand in terms of quantities and qualities. The Saline Water Conversion Corporation (SWCC) was established as a ministerial agency under the

I thank the Research Institute of King Fahd University of Petroleum and Minerals for the support provided to complete this study.

MAW in 1965, and then as an independent corporation within the MAW in 1974, to be responsible for construction, operation, and maintenance of desalination plants for drinking water production. The Water and Wastewater Authority (WWA) is an independent agency under the Ministry of Rural and Municipal Affairs to distribute drinking water and to collect and treat wastewater in different cities and towns of the kingdom.

More recently, the government has modified the past approach of increasing supplies to meet rising demand. To protect the community of interest that constitutes the traditional basis of Islamic customary water law, the government has taken several measures to protect the sustainability of aquifer systems and groundwater resources. Laws, regulations, and *fatwa* were developed, in accordance with Islamic law, to deal with water management issues, including measures to reduce national water demand and augment available water resources.

According to Islamic law and custom, water is to be used first for domestic purposes, then for animals, and lastly for agriculture. The Prophet (pbuh) mentions, among those "whom Allah will neither talk to, nor look at, on the Day of Resurrection:... A man who withholds his superfluous water. Allah will say to him, 'Today I will withhold My Grace from you as you withheld the superfluity of what you had not created.'"[2] Regarding animals, Allah's Apostle said, "He who digs a well in the desert when there is pasture around this well and when there is no other water nearby cannot prevent the animals from slaking their thirst at this well."[3] and "One should not prevent others from watering their animals with the surplus of his water in order to prevent them from benefiting by the surplus of grass."[4] In Saudi Arabia, industrial and recreational uses come fourth and fifth, respectively. The order of the last two purposes was ranked according to the application of Islamic customs in the country and through reasoning rather than from strict doctrine. This chapter describes Saudi Arabia's available water resources, and how demand is managed for different purposes according to Islamic law.

Available water resources

The Kingdom of Saudi Arabia has an area of about 2.25 million km^2, most of which is located in arid regions. The available surface water and groundwater resources are limited, precipitation rates are low, and evaporation is high. The average annual rainfall is less than 150 mm in most of the country. During the last two decades, the kingdom has experienced comprehensive development in all sectors coupled with high growth rates in population and living standards. The annual national water

Table 1 **Growth of Water Use in Saudi Arabia, 1980–2010**
(millions of cubic metres)

	Domestic and industrial	(per cent)	Agricultural	(per cent)	Total
1980	502	21.3	1,850	78.7	2,352
1990	1,650	6.06	25,589	93.94	27,239
1992	1,870	5.9	29,826	94.1	31,696
1997	2,063	11.17	16,406	88.83	18,469
2000	2,900	20.57	11,200	79.43	14,100
2010	3,600	19.67	14,700	80.33	18,300

Sources: MOP 1990; Dabbagh and Abderrahman 1997 (agricultural and total use, 1990 and 1992).

demand has increased from 2,352 million cubic metres (MCM) in 1980 to about 27,239 MCM in 1990, and to more than 30,000 MCM in 1992 (table 1).

Conventional resources

The annual runoff in the kingdom is estimated to be about 2,230 MCM. There are 185 dams, with a total storage capacity of 775 MCM, for groundwater recharge and flood control.

Groundwater is stored in more than twenty layered principal and secondary aquifers of different geological ages (MAW 1984), with groundwater quality varying between sites and aquifers. Isotopic analyses show that the fossil groundwater in these aquifers is ten to thirty-two thousand years old. The estimated groundwater reserves to a depth of three hundred metres below ground surface are about 2,185 billion cubic metres with a total annual recharge of 2,762 MCM (Al Alawi and Abdulrazzak 1994; Dabbagh and Abderrahman 1997). The renewable groundwater resources are mainly stored in shallow alluvial aquifers and in basalt layers of varying thickness and width, which are found mostly in the southwest. These aquifers store about 84 billion cubic metres with an average annual recharge of 1,196 MCM.

Non-conventional resources

Thirty-five desalination plants have been built, at a cost of about US$10 billion, to produce potable water from sea water and raw groundwater along the Red Sea coast and the Arabian Gulf coast using the multistage flush system and reverse osmosis (Bushnak 1997, 93). At present, Saudi

Table 2 **Water supply in Saudi Arabia, 1990–1997**
(millions of cubic metres)

	1990	(per cent)[a]	1992	(per cent)[a]	1997	(per cent)[a]
Surface water and shallow aquifers (renewable)	2,100	13	2,140	7	2,140	12
Groundwater (non-renewable)	24,489	83	28,576	90	15,376	83
Desalination	540	3	795	2	795	4
Treated wastewater effluents	110	0.7	185	0.6	185	1
Total	27,239		31,696		18,496	

Sources: MOP 1990 estimate; Dabbagh and Abderrahman 1997 (1992 total).
[a] Percentages do not add to 100 due to rounding.

Arabia is the largest producer of desalinized water in the world. Annual water production has reached about 795 MCM and annual capacity will reach about 1,050 MCM by 2000.

Desalination unit cost is about US$0.70 or SR 2.6 per cubic metre (US$1 = 3.751 Saudi riyals, SR) for large-size desalination plant (Bushnak 1997, 93). About SR 3–4 should be added to the total water costs for transporting desalinated water to cities and towns. Thus, 1 cubic metre of desalinized water delivered to a house costs about SR 5.5–6.6.

In addition, it is estimated that about 1,000 MCM of wastewater were generated in the country in 1996, and this is expected to increase to about 1,500 MCM by the year 2,000 (Ishaq and Khan 1997). About 41 per cent of municipal wastewater is treated, and in 1997, about 185 MCM or 18.5 per cent of the treated wastewater was recycled for irrigating agricultural crops and landscape plants and for use in refineries.

Domestic water demand management

As discussed in other chapters in this volume, including those by Amery and by Kadouri et al., in order to prevent scarcity of water or to prevent it being controlled by one person, the Prophet (pbuh) endeavoured give to all people the right to water. This principle is particularly important for Saudi Arabia, with its limited resources and rapidly growing demand. The total population of Saudi Arabia has increased from about 7.7 million in 1970 to 11.8 million in 1990 and is expected to reach 19 million in 2010, if the present growth rate of 3 per cent per annum continues. Consequently, domestic water demand has increased from about 446 MCM in 1980 to about 1,563 MCM in 1997, and is expected to reach 2,800 MCM in 2010 (Al-Alawi and Abdulrazzak 1994; Al-Tukhais 1997). Hence the large-scale construction of desalination plants, which at present supply 46 per cent of domestic demand.

As noted in the chapters by Shah and by Atallah et al. in this volume, however, water conservation was also emphasized during the early times of Islam, and to reduce the domestic water demand in Saudi Arabia, several water control and conservation measures have been introduced. These include the following.

- In 1994, water tariffs were introduced to enhance the people's awareness of the value of water production. The tariff per cubic metre of potable water is US$0.04 (SR 0.15) for the first one hundred cubic metres, US$0.27 (SR 1.0) for the second one hundred cubic metres, US$0.53 (SR 2.0) for the third one hundred cubic metres, and US$1.07 (SR 4.0) for the fourth one hundred cubic metres. The water charges for a medium-sized middle-class family (six persons) living in a small house with garden (assuming water consumption of about 200 cubic metres a month), and with an average income of SR 4,000 a month, are less than SR 200 a month (US$55 a month). However, the charge for water is only a small fraction of the actual cost of water production and transportation, which ranges between about SR 1,120 and SR 1,320.
- Leakage control measures have been implemented to minimize water losses from water supply networks.
- Treated wastewater recycling has been implemented; for example, ablution water is recycled for toilet flushing at the two Holy Mosques at Makka and Al-Medina Al-Monawwarah.
- Highly saline water from Wadi Malakan near Makka is used instead of desalination water for toilet flushing at the Holy Mosque at Makka.

Irrigation water demand management

The cultivated area in the kingdom has increased from less than 0.4 million ha in 1971 to 1.62 million ha in 1992 (MAW 1992), and total consumption of irrigation water has increased from about 1,850 MCM in 1980 to 29,826 MCM in 1992 (table 1). The threshold increase in the agricultural area started after 1979. Because of its responsibility for making water available to the people for different uses, including irrigation as the third priority, the government gave financial support to farmers for well drilling and the introduction of modern and efficient irrigation systems. Extension services were also introduced to help farmers in proper scheduling of irrigation water to avoid excessive use. A preliminary assessment of the cost of water production for irrigation from wells with depths of less than four hundred metres was between SR 0.20 and SR 0.50 for large irrigation schemes.

Table 2 shows that non-renewable groundwater from shallow and deep aquifers supplied about 28,576 MCM in 1992 for irrigation use. This represented about 94 per cent of the total irrigation water use and 90 per cent of the total national water use in Saudi Arabia. The total number of drilled wells increased from about 26,000 in 1982 to about 52,500 in 1990, and hundreds or even thousands of production wells were thickly clustered in some agricultural areas. In several agricultural regions, excessive water pumping has resulted in negative effects on groundwater levels and on quality. Consequently, improvement of groundwater management and reduction in irrigation water consumption, especially for wheat cultivation, became essential for maintaining the long-term productivity and quality of the aquifers. Understanding this serious issue, the government, after consultation with leading Islamic scholars and with specialists in agriculture, economics, and water, took several measures and developed regulations to improve the management of water demand and to protect and conserve water resources.

Regulation of well drilling

"The Prophet said, 'A Muslim is the one who avoids harming Muslims with his tongue and hands. And a Muhajir (emigrant) is the one who gives up (abandons) all what Allah has forbidden.'"[5] This means that the Muslim is forbidden by Islamic law to cause any harm to others including his community. Furthermore, the Prophet (pbuh) recognized that the ownership of wells or any other water source requires the ownership of a certain extent of bordering land or *harim* on which it was forbidden to dig a new well. This was to avoid any negative effects on the quality and abundance of the well.

Following this general Islamic trend, a royal decree was issued in 1980 to regulate well drilling and to protect aquifers from exploitation and pollution. Special permits must be issued in advance by MAW to drill or deepen any well, and drilling and deepening must follow approved designs and be carried out and under supervision of MAW. Well owners and the drilling companies face penalties for not observing this decree.

Reduction in wheat price supports

Saudi Arabia's largest crop is wheat, with a total of 907,309 ha or 56 per cent of the total cultivated area in 1992, while fodder crops, vegetables, and fruits accounted for 18, 7, and 6 per cent of the total agricultural area respectively. The wheat production of 4.25 million tons in 1992 far exceeded the predicted national demand of 1.22 million tons (MOP 1990), which hindered diversification of agricultural production and resulted in unnecessary consumption of large volumes of groundwater: in 1992, irrigation demand of wheat was 9,895 MCM or 33 per cent of the total national irrigation water consumption.

In 1993, the government reduced the area of wheat cultivation eligible for price support to 25 per cent of its previous size. This was to reduce wheat production to the level of the annual consumption, encourage farmers to diversify crop production, and reduce irrigation water consumption. The reduction in water use was projected to amount to about 7,400 MCM per year or 25 per cent (assuming a 75 per cent reduction in wheat area). In actual fact, the area under wheat dropped by about 325,000 ha between 1992 and 1994, and as shown in table 2, although the water supply from other sources remained constant from 1992 to 1997, the reduction in wheat production subsidies resulted in a drop from 28,576 MCM to 15,376 MCM in water pumped from non-renewable aquifers. This reduction positively affected groundwater levels and quality in different wheat areas in the kingdom. Field measurements of groundwater levels in deep observation wells in a large irrigation scheme in the Eastern Province have shown a recovery after reduction of the area under wheat of about 20–30 per cent from the drawdown recorded in previous years. Recently, the MAW announced similar positive effects on groundwater levels in other regions of the kingdom as a result of reductions in wheat cultivation.

Reuse of wastewater effluents for irrigation

Millions of cubic metres of wastewater effluents used to be produced and disposed of without reuse. This was not for technical reasons, but because it was not clear if the effluents were pure according to Islamic views, even after removal of impurities by proper treatment. After lengthy and deep investigations and discussions with scientists and specialists, a special *fatwa* on the matter was issued by the Council of Leading Islamic Scholars (CLIS) of Saudi Arabia in 1978. The *fatwa* (CLIS 1978) postulated that

> Impure waste water can be considered as pure water and similar to the original pure water, if its treatment using advanced technical procedures is capable of removing its impurities with regard to taste, colour and smell, as witnessed by honest, specialized and knowledgeable experts. Then it can be used to remove body impurities and for purifying, even for drinking. If there are negative impacts from

its direct use on the human health, then it is better to avoid its use, not because it is impure but to avoid harming the human beings. The CLIS prefers to avoid using it for drinking (as possible) to protect health and not to contradict with human habits.

This *fatwa* demonstrates the dynamic nature and wisdom of Islamic law when confronting the changing needs of the Muslim community. It was an important step toward the reuse of wastewater effluents for different purposes depending on its degree of treatment, such as drinking, ablution, removal of impurities, and restricted and non-restricted irrigation. At present, about nine thousand hectare of date palms and forage crops near Riyadh are irrigated using about 146 MCM of wastewater effluents. Wastewater is also reused for irrigating landscape plants, trees, and grass in municipal parks in several cities, such as Dhahran, Jeddah, Jubail, Riyadh, and Taif.

Other water demand reduction measures

The MAW has considered the introduction of water meters on farm pumps to help in minimizing overpumping and water losses. There is also the possibility of shifting of some fodder and cereal cultivation from zones of high irrigation water consumption to areas of lower consumption, thus saving considerable quantities of irrigation water. The MAW is active in improving public knowledge of the value of water conservation in the news media and in educational institutions.

Industrial water demand management

Although industrial water constitutes only a small portion of total demand, certain industries require special water qualities; and the environmental effects of mismanaging industrial wastewater represent a major hazard. Industrial water demand increased from about 56 MCM per year in 1980 to 192 MCM in 1990, and is expected to grow to about 500 MCM in 2010 (Al-Alawi and Abdulrazzak 1994, and author's estimate). The growing demand is satisfied mainly by costly desalination in some industries, especially food, although groundwater satisfies other types of industries. Industrial demand varies among regions of the kingdom. In some industrial plants, part of the effluent is recycled. However, uncontrolled disposal of wastewater has had negative effects on the environment and groundwater.

The following legislation and measures have been taken to improve industrial water demand management.

- To minimize industrial water demand, to maximize wastewater recycling, and to protect the environment, the government has established large industrial cities in different parts of the kingdom. Each city contains tens or hundreds of factories. Industrial wastewater is collected, treated, and recycled within each city at the plant level for industrial and landscape purposes. The industrial cities have specifications for the quality of the wastewater collected from factories.
- Closed water cycles have been introduced in industrial plants to minimize wastewater disposal, reduce groundwater pumping, and protect the environment. In this approach, wastewater is converted into good quality condensate by evaporation at low temperature under vacuum. This technology was introduced to large industrial plants in 1995 (Abderrahman 1997).

Conclusions

The Kingdom of Saudi Arabia has followed Islamic law, or *sharia*, in all aspects of life since its establishment about a hundred years ago. *Sharia* considers water as the main component of the sustainability of a nation's life and security. The government realized the problems created by the combination of arid climatic conditions, limited water supplies, and rapid growth in water demand, and introduced specialized agencies for water production and distribution, as well as regulations, measures, and *fatwa* in agreement with Islamic law to achieve effective water demand management in the interests of the community and its natural resources. This was implemented in continuous consultation with leading Islamic scholars and specialists in water, agriculture, planning, and economics. Examples of the regulations are the reduction in domestic water demand by the introduction of new water pricing policies, leakage detection and control measures, and promotion of public awareness of the value of water. Another example is the reduction of support to wheat production, which has resulted in a reduction in irrigation water demand of about 25 per cent. Furthermore, the *fatwa* permitting reuse of wastewater effluents especially for irrigation has resulted in the reuse of millions of cubic metres of treated effluent every year for this purpose. Recycling of treated industrial wastewater has been encouraged, and has been implemented by various industrial plants. Other measures have been introduced to control well drilling and to monitor water consumption at the farm level so as to avoid overpumping and to protect aquifers. Farmers are also encouraged to use modern irrigation systems and to adopt irrigation scheduling to minimize water demand. In all these ways, the experience of Saudi

Arabia in using Islamic principles in water demand management has been successful in satisfying growing water needs and protecting water resources. This demonstrates that Islamic regulations are sufficiently dynamic, flexible, and reasonable to solve the challenges faced by Muslim nations in such vital issues as water.

Notes

1. Al-Bukhari 2.18.
2. Al-Bukhari 3.557.
3. Al-Bukhari 5550, in *Hadith Encyclopedia*.
4. Al-Bukhari 9.92.
5. Al-Bukhari 1.9.

REFERENCES

Abderrahman, W. A. (1997), "The Use of Closed Water Cycle in Industrial Plants in Saudi Arabia," in *Proceedings of the Conference on Development and Environmental Impact, 21–23 September*, Ministry of Municipal and Rural Affairs, Riyadh.

Al Alawi, M. and Abdulrazzak, M. (1994), "Water in the Arabian Peninsula: Problems and Perspectives," in P. Rogers and P. Lydon (eds.), *Water in the Arab World: Perspectives and Progress*, Division of Applied Sciences, Harvard University, Cambridge, Mass.

Al-Tukhais, A. S. (1997), "Water Resources and Agricultural Production in Saudi Arabia: Present and Future," in *Water Resources and Its Utilization in Saudi Arabia: Proceedings of the First Saudi Conference on Agricultural Sciences, 25–27 March 1997*, College of Agriculture, King Saud University, Riyadh.

Bushnak, A. (1997), "Water Desalination and Wastewater Reuse: Review of the Technology, Economics and Applications in the ESCWA Region," *Expert Group Meeting on Development of Non-Conventional Water Resources and Appropriate Technologies for Groundwater Management in the ESCWA Member Countries, 27–30 October, Manama, Bahrain*. Economic and Social Commission of Western Asia (ESCWA)/UN, Beirut.

CLIS (Council of Leading Islamic Scholars) (1978), "Judgement Regarding Purifying Wastewater: Judgement no. 64 on 25 Shawwal, 1398 AH, Thirteenth Meeting of the Council of Leading Islamic Scholars (CLIS) during the Second Half of the Arabic month of Shawwal, 1398 AH (1998)," *Taif: Journal of Islamic Research* 17, pp. 40–41.

Dabbagh, A. E. and Abderrahman, W. A. (1997), "Management of Groundwater Resources under Various Irrigation Water Use Scenarios in Saudi Arabia," *Arabian Journal of Science and Engineering* 22 (special theme issue on water resources in the Arabian Peninsula), pp. 47–64.

Ishaq, A. M. and Khan, A. A. (1997), "Recharge of Aquifers with Reclaimed Wastewater: A Case for Saudi Arabia," *Arabian Journal for Science and Engineering* 22, pp. 133–41 (special theme issue on water resources in the Arabian Peninsula).

MAW (Ministry of Agriculture and Water) (1984), *Water Atlas of Saudi Arabia*, MAW, Riyadh.

——— (1992), *Agricultural Statistical Year Book*, vol. 7, MAW, Department of Economic Studies and Statistics, Riyadh.

MOP (Ministry of Planning) (1990), *Fifth Development Plan*, MOP, Riyadh.

7

Sociocultural acceptability of wastewater reuse in Palestine

Nader Al Khateeb

The concept of using wastewater for irrigation can be traced back more than two thousand years, when crops in Greece were irrigated with such effluent. Despite this, the present widespread interest in the concept has developed only recently. The use of treated wastewater has increased significantly in developing countries since about 1980. This is partly because increasing population and increasing consumption of water per capita have meant that more and more wastewater is being produced in urban and rural areas of developing countries. In addition, increasing numbers of sewage treatment works have been built, and are in the process of being built, in developing countries – the result of the United Nations (UN) designating the 1980s as the International Water Supply and Sanitation Decade, following the UN Water Conference. At the same time, the increasing scarcity of water in many arid and semi-arid countries has led planners to search for additional sources of water that can be used economically and effectively to promote further development. Wastewater is an obvious candidate, and has come to be considered as a new and unconventional source of water for agricultural production, which can contribute to the alleviation of hunger in many countries. Finally, increasing interest in environmental and health issues in developing countries has led to interest in safe and beneficial disposal of waste water.

In many sociocultural environments, however, the use of treated wastewater raises the question: Is this new source of water culturally acceptable? This is not a question that can always be answered simply. Cultures are rarely homogeneous and frequently contain a complex variety of subcultures with widely differing orientations. Furthermore, cultures are not fixed entities: values, beliefs, and customs change and can be made to change. In addition, the most appropriate methods for gathering data, above all those of social anthropological fieldwork, have rarely been used in this culturally sensitive area.

All these considerations are of obvious relevance to Palestine, where a pilot demonstration project was recently implemented in the town of Nablus. In this project, tricking filters and activated sludge systems were used to treat sewage. Eggplants, pepper, apples, grapes, and peaches were grown and irrigated with effluent, and some also with freshwater as a control. The crops were tested in the laboratories of the Palestinian Ministry of Agriculture and shown to be safe to eat. In addition, we established (mainly on the basis of the *fatwa* of the Council of Leading Islamic Scholars of Saudi Arabia, discussed by Abderrahman in this volume) that Islam permits the reuse of wastewater for irrigation and other purposes – an important matter, since religion can act as a promoter or inhibitor of new ideas, and most Palestinians are Muslims. Having established this point, we carried out a survey to determine the social acceptability of wastewater reuse among the Palestinian general public and farmers. Questionnaires were designed, and two engineers were contracted to select random samples in every district, interview those selected, and fill out the questionnaires. In some cases, the questionnaires were given to respondents to fill out themselves and return them the following day. In addition, visitors to the demonstration site were interviewed. The survey was mainly intended to assess the sociocultural perspective for wastewater reuse so as to mitigate water shortages, and to assess the possibilities of increasing wastewater reuse through public awareness campaigns. The responses from the general public, farmers, and visitors to the Nablus project are summarized below.

The general public

A random sample of 480 people from Bethlehem, Hebron, Jenin, Jericho, Nablus, Qalqilya, Ramallah, and Tulkarem districts was surveyed to determine their responses to the concept of reusing treated wastewater in agricultural irrigation. The following results were obtained.

- About 88 per cent of those surveyed believe there is a water shortage in Palestine.
- Although 73 per cent have piped water systems, 57 per cent still use household percolation pits to dispose of their seepage.

- Almost all the respondents (85 per cent) believe that reusing wastewater is one option for coping with the shortage of irrigation water.
- Over half of those surveyed (55 per cent) believe that wastewater is no longer simply waste, but a usable water source.
- More than 78 per cent believe that there is an acute need to reuse wastewater for irrigation.
- Over 80 per cent said they are willing to practice wastewater reuse.
- About 50 per cent felt that the reuse should be to irrigate crops that can be exported while the other 50 per cent said the irrigated crops should be marketed locally.
- About 65 per cent of respondents said they were willing to buy crops irrigated with treated waste water.
- Nearly 80 per cent had no knowledge of how to deal with wastewater, and 94 per cent said it was necessary to take precautions when dealing with wastewater.
- About 93 per cent said they would accept having wastewater treatment plants in their towns.
- Nearly 80 per cent of those surveyed had never seen a wastewater treatment plant.
- Nearly 60 per cent think that drip irrigation is the most appropriate method for irrigation with treated wastewater.
- About 65 per cent of respondents said that diarrhoea was the most prevalent disease in their towns.
- Over 66 per cent said their family income was less than two thousand new Israeli shekels per month (US$1 = NIS4.1).
- About 50 per cent said they preferred getting information about wastewater reuse through television, while the rest preferred getting information through radio or newspaper reports.
- Almost all the respondents believed that wastewater use is allowed by Islam under conditions that prevent harm to the users.

Farmers

A random sample of seventy farmers from the villages of Deir Al Ghosoun in Tulkarem District, Beit Eiba in Nablus District, and Taffouh in Hebron District were interviewed to determine their response to the concept of reusing wastewater in agricultural irrigation.

- Most of the respondents said they had heard about wastewater reuse in irrigation.
- Almost all believed that it is technically and economically possible to safely reuse wastewater to address irrigation water shortages.
- Many respondents said that the main obstacle they faced was the lack of sufficient water and the high cost of fertilizer.

- The majority of farmers said that wastewater reuse is not prohibited religiously and can be practised.
- Most of the farmers said that wastewater has an advantage over freshwater for irrigation because it contains most of the nutrients required by plants.
- The farmers are ready to accept the reuse of wastewater for irrigation on their farms if the effluent quality is assured, and the law allows it, and they do not have to change their cropping patterns.
- More than 53 per cent of the farmers said they were willing to pay up to NIS 1 per cubic metre of treated wastewater and a further 39 per cent said they were willing to reuse wastewater if it is provided free of charge.
- Almost 65 per cent of the farmers said they did not know enough about sludge from sewage treatment plants, but that they were willing to use it as a fertilizer if it is safe and not contaminated with pathogens.
- About 76 per cent of the farmers said they were willing to pay for the sludge to use it as a fertilizer if it is safe.

Demonstration site visitors

In addition to the survey, many people were invited to the demonstration site, including trainees from various municipalities in the West Bank who were taking courses at Bir Zeit University on wastewater collection, treatment, and reuse (funded by the German government). All of the visitors were briefed on the concept of wastewater treatment and reuse, and the benefits and risks. Interviews with the visitors produced the following results.

- Most of the visitors had not seen a wastewater treatment plant before. This was reflected in their views on the possibilities of recycling the wastewater.
- When the visitors had seen the demonstration project, they became supportive of the idea of wastewater reuse.
- The visitors were concerned about the health and smell aspects of wastewater treatment.
- Most of the visitors were in favour of using wastewater on trees, or on vegetables that are eaten cooked.
- The visitors believe that there is a water shortage in the area, that wastewater could be reused in agriculture, and that more freshwater could then be allocated for domestic use.
- The visitors suggested that widespread awareness campaigns should be implemented, and stressed the need for more information about

wastewater treatment and about the results that could be obtained from projects such as that at Nablus.

- The visitors were satisfied with the concept of establishing a demonstration project as a first step before constructing large-scale wastewater treatment plants.
- The visitors supported the concept of involving local experience in the planning and decision-making for wastewater treatment and reuse.
- The visitors said that the polluters should be responsible for the treatment of their wastewater and not the farmers.

Conclusions

Based on the sociocultural survey of the farmers and the public in the West Bank, we can conclude the following:

- Palestinians in the West Bank believe that wastewater reuse is acceptable in Islam providing that the effluent quality is safe and does not harm the health of the users.
- Palestinians in the West Bank believe that there is a water shortage in Palestine and (treated) wastewater can be safely reused in irrigation to conserve more freshwater.
- Most of those surveyed had never seen or experienced wastewater treatment, and there is a need to intensify demonstration activities.
- Most respondents think that the reuse of raw sewage is dangerous whereas treated sewage can be a useful water resource.
- The respondents are willing to consume products irrigated with treated wastewater.
- Those surveyed prefer getting information about wastewater treatment and reuse through television.
- There is a need to initiate a national campaign on possible options for wastewater reuse. Such a campaign should include visits to the demonstration site.

FURTHER READING

Al Khateeb, N. (1997), *Nablus Wastewater Treatment and Reuse Demonstration Project: Final Report.* United Nations Development Programme, Jerusalem.

Al Khateeb, N., Assaf, K., Kally, E., and Shuval, H. (1993), *A Proposal for the Development of a Regional Water Master Plan*, Israel-Palestine Centre for Research and Information, Jerusalem.

Al Yaazigi, N. and Dadah, J. (1994), *The Utilisation of Dry Sludge as Fertiliser in Gaza Strip*, Government Education College, Gaza.

DANIDA (Danish International Development Assistance) (1996), *Urgent Action Plan for Wastewater Management: Gaza Governorates.* DANIDA, Gaza.

Gearheart, R., Bahri, A., and Al Hamaidi, M. (1994), *Wastewater Treatment and Reuse Strategy for Gaza and the West Bank: Water and Wastewater Sector*, Palestinian Economic Council for Development and Reconstruction, Jerusalem.

IWACo-Euroconsult (1995), *Gaza Environmental Profile – Water Reuse, Prefeasibility Study*, Palestinian Environmental Protection Agency, Gaza.

Juanio, M. and Amiel, A. (1992), *Impact of Sewage on Groundwater Quality in the Gaza Strip: Final Report*, Water Commission, Gaza Israeli Civil Administration and the Technical Research and Development Foundation, Gaza.

Nashashibi, M. (1995), "Wastewater Treatment Strategies in Palestine," M.Sc. thesis, Delft.

PWA (Palestinian Water Authority) (1996), *Wastewater Reuse in Agriculture in Gaza Governorates*, PWA, Gaza.

Sourani, G. (1991), "Desalination: A Feasibility Study for the Gaza Strip," M.Sc. thesis, Loughborough University of Technology.

TAHAL Consulting Engineers (1993), *Recycling of Wastewater for Environmental Protection and Water Supply in Agriculture*, TAHAL Consulting Engineers Ltd., Tel Aviv.

World Bank (1994), *Strategic Study on Wastewater Reuse*, World Bank, Washington, D.C.

WHO (World Health Organization) (1973), *Reuse of Effluent: Methods of Wastewater Treatment and Health Safeguards*, Technical Report 517, WHO, Geneva.

8

Water rights and water trade: An Islamic perspective

M. T. Kadouri, Y. Djebbar, and M. Nehdi

The 1992 United Nations Conference on Environment and Development held in Rio de Janeiro, and the International Conference on Water and Environment held in Dublin the same year, generated a consensus that the developing countries face two great challenges in the water supply and sanitation sector. The first challenge is to complete the old agenda of providing household service (Bronsro 1998). Although significant progress has been made, much remains to be done. During the 1980s, the number of people without safe water decreased from 1.8 to 1.2 billion, while the number of people without suitable sanitation remained static at about 1.7 billion. Despite these relative successes, the shortfall in adequate water supply and sewerage has dire consequences for human development (Serage El-Din 1994). The second challenge is the new agenda of sustainable development. This challenge includes the need for long-term, more efficient, and equitable water supply.

To address the water crisis in the Middle East described in the introduction to this volume, the use of water demand management tools such as pricing, regulations, technology, and education is imperative. Although these tools have the potential to help many water utilities move out of the current crisis while simultaneously extending and improving services, realization of these benefits is yet to be achieved because the history of demand management in developing countries is short.

Water pricing through user-paid fees is one of the most controversial water demand management tools. It requires better market-like instruments to promote sustainability, raise revenues, and provide fair prices, particularly for the poor. Implementing water pricing as a water demand management tool requires not only an understanding of the full spectrum of urban issues, but also institutions to ensure that those charged with reforms are able and willing to carry them out.

Islam plays a pivotal role in all aspects of life in the Middle East, from major legislation to elementary social behaviour, and any solution to water management problems must consider the Islamic reality of the region. This chapter addresses water rights and water pricing in Islam, and outlines an Islamic water management perspective in the Middle East.

Economic theory, water trade, and price distortion

The limited water supply from currently available sources is triggering development of new, more distant, and more expensive water sources (World Bank 1993). Typically, marginal costs for new water sources are high: for example, new water sources in Algeria and Egypt will cost two to three times more than existing ones (World Bank 1992).

Economic theory indicates that water should be priced at the marginal cost of providing the next increment of water resource. However, water historically has been provided below cost or even free. Most often, water tariffs are set not to recover marginal costs, but at average financial costs or even less. The difference essentially is that cost recovery pricing reflects past costs, whereas marginal-cost pricing reflects future costs. Average-cost pricing, therefore, can encourage overuse. Furthermore, in many places, flat-rate or declining-block-rate structures are still being used, and these offer no incentives to conserve water (Bronsro 1998). Bronsro (1998) proposes that the positive assumption of universal water supply should be reconsidered and that there should be a move toward economic pricing, which will likely involve substantial tariff increases in water-stressed cities.

A premise of sustainability is that water is a scarce economic resource. As such, the price of water should include not only direct costs such as those of transporting it, but also the external costs of environmental degradation, and the opportunity costs of uses forgone. In any case, calculating environmental costs is likely to be controversial at best, because the benefits of aquatic systems, such as habitat for fish, animals, and plants; climate moderation; and aesthetic value are not traded in the marketplace (Bronsro 1998). The prices charged for water extraction are unlikely to include these values.

In a modern context, effective management has to handle collection, treatment, and distribution of water, as well as maintenance of water resources and infrastructures. The latter have a cost that can only be recovered by effective pricing. Calculating prices is often easier than collecting tariffs. Powerful sociopolitical forces may agitate against raising tariffs for an essential commodity such as water. If an influential minority benefits from water subsidies at the expense of the majority, the subsidy can be hard to abolish. When prices are distorted, markets are non-competitive, and there is private trading in public goods, economists say that markets have "failed" (Panayotou 1993). Thus, in Jakarta roughly 20 per cent of the city's eight million people receive piped water from municipal water connections. The rest depend on private wells or purchase water from various private vendors. Well water is inexpensive, but is contaminated and unsafe. Its overuse leads to public health problems, aquifer depletion, and land subsidence (Bronsro 1998). Crane (1994) reported that those who did not have access to piped water in Jakarta paid six to fourteen times more than those who did. Such price variations are typical: Bahl and Linn (1992) summarized vendor prices for water relative to municipal charges in various countries – in Burkina Faso three to five times higher, in Ghana thirteen to twenty-five times, in Kenya (Nairobi) seven to ten times, and in Uganda (Kampala) four to ten times. Not surprisingly, in Jakarta, those who paid higher prices purchased much less water than those who paid lower prices – fourteen litres per day per person versus sixty-two litres. The higher-price consumers generally limited their meagre ration of purchased water to drinking and cooking, and used well water for other purposes.

Although benefiting only a minority, such a market structure creates obvious problems for the environment, the water utility, and the average customer. The standpipe vendors purchase water from the municipal system and resell it at a substantial markup. Entry to this market is controlled, while prices effectively are not. Therefore, prices are high and the vendors extract monopoly rents. As indicated earlier, Islam forbids such conduct, which creates vested interests and resists price reforms. In contrast, access to groundwater is not controlled, and this leads to overuse. Many of the poor face a choice between high prices and poor water. In theory, they would be better off with house connections, because the average vendor customer could increase his water consumption fivefold, while still decreasing his water bill. However, many are prevented from connecting to the municipal system by ignorance about the options for water supply, as well as by credit constraints, bureaucratic obstacles, and official corruption (Lovei and Whittington 1993; Crane 1994). A further problem for the city supplier is that the proportion of water lost to leakage and theft exceeds 50 per cent (Bhattia and Falkenmark 1993).

Pricing of water in Islam

Before discussing water rights and pricing from an Islamic perspective, the concept of wealth ownership in Islamic jurisprudence must first be understood. Property in Islam is a social function, that is, wealth belongs to God and a person simply assumes a managerial position to increase wealth and use it properly. The word "wealth" ("ma-li" in the Arabic language) has no significance on its own; it is merely a relationship. This meaning is emphasized by the Quran: "*O ye who believe! Spend out of (the bounties) we have provided for you*"[1] says that "Indeed, the wealth that was bestowed upon you belongs to Allah for He has created it. He has only designated you as gerents and allowed you to enjoy it."

However, this should not be taken to mean that Islam jeopardizes economic incentives by "externalizing" property. It basically balances private incentives with social optimality. The economic concept in Islam is based upon reward: a person should be rewarded for his work, and work is most honoured. The Prophet (pbuh) says: "If anyone revives dead land, it belongs to him...."[2] Market incentives should drive the economy and a government should not interfere in the market except to prevent unfair competition, and to inhibit illicit (*haraam*) practices. Muslim scholars agreed that Islam does not allow the government to fix prices for goods, including water – the market itself fixes the prices. It was reported that when some people complained to the Prophet about high prices and asked him to adjust them, he refused to do so and said that "Allah is the one Who fixes prices, Who withholds, gives lavishly and provides, and I hope that when I meet Allah, none of you will have any claim on me for an injustice regarding blood or property."[3] This indicates that in Islamic law, under normal circumstances prices should not be fixed. As will be shown later, however, there are exceptions to this rule.

The advantage of the dissociation between God's fundamental ownership of wealth and humanity's "managerial" ownership is twofold: first, one has no right to harm himself, his belongings, others, or the environment; and second, one cannot abuse the sources of wealth or put one's own individual interest ahead of the public interest in conducting affairs. Islam promotes moral self-regulation to enhance social justice and to combat corruption, then sets forward a system of law to enforce its moral code.

A fundamental principle in dealing with wealth-generating resources in Islam is combatting unfair distribution, "*in order that it may not (merely) make a circuit between the wealthy among you.*"[4] Thus, Islamic jurisprudence attempts to balance the reward of work and the public interest in managing water resources. It is reported that the Prophet said, "Muslims have common share in three (things): grass, water and fire."[5] The Prophet discouraged the selling of water. Amrou Ibn Dinar said, "We do not know whether he meant flowing water in nature (in rivers and lakes) or transported water (with added value)." However, most Muslim scholars (Zouhaili 1992) agree that water could be sold like any other commodity. The Prophet once said: "he who purchases the Ruma Well and offers its water to Muslims free of charges will be granted paradise."[6] This saying indicates that wells can be traded, and so can their water. He also said: "it is better for anyone of you to take a rope and cut the wood (from the forest) and carry it over his back and sell it (as a means of earning his living) rather than to ask a person for something and that person may give him or not."[7] Thus, Muslim scholars conclude that water, like lumber and other public commodities, could be sold and traded (Zouhaili 1992).

More specifically, most Muslim scholars subdivide water resources for trading purposes into three categories (Sabeq 1981; Zouhaili 1992): private goods, restricted public goods, and public goods.

Water stored in private containers, private distribution systems, and reservoirs is considered as a private good. This also includes water that has been extracted from wells and rivers using special equipment or obtained through water distribution companies. This water belongs to its owner and cannot be used without his permission. The owner has the right to use it, trade it, sell it, or donate it. Even though this water is private, a person in need can use it after asking for the owner's permission. Likewise, treated water can be traded because the organization responsible for the treatment has spent money and invested work in it (added value or reward for work). This ruling can encompass water from treatment plants, water privately transported and stored, and any water to obtain which work, infrastructure, and knowledge have been invested.

Water bodies such as lakes, water streams, and springs that are located on private lands are considered to be restricted public goods. This water does not belong to its owner in the large sense of ownership; rather, the owner merely has special rights and privileges over other users. For instance, other users can use this water for drinking and basic needs, but they cannot use it for agricultural and industrial purposes without the permission of the owner. However, the *Shafii* believe that whoever digs a well owns its water, which is therefore considered to belong to the first category – private goods.

Water in rivers, lakes, glaciers, aquifers, and seas, and from snow and rainfall is a public good. Anyone has the right to use it (properly) for drinking and for agricultural and industrial purposes as long as this does not hinder environmental or public welfare. This water can be transported in pipes, canals, and containers for private use. The government should not prevent its use, unless it can prove that the use will produce harm to the public welfare, damage to the environment, overuse, or unfair trade. Water falling in this category cannot be sold or bought for private interest (Zouhaili 1992). However, if any value is added, such as treatment, storage, and transportation, the water becomes a private good, and it can be sold to recover cost and generate profit.

Although Islamic jurisprudence does not go into the specifics of setting static regulations for pricing or market control, it does put forward a set of general principles that guide the pricing of any traded goods including water. These guiding principles can be summarized as follows (Sabeq 1981; Zouhaili 1992).

- In the spirit of the Quran and the Prophet's sayings, Muslim scholars encourage giving water away free of charge, indicating that Allah will reward those who do so. However, they indicate that the owner of private water should not be forced to provide water free of charge except in compelling conditions, and where other sources of water are not available. Even in such conditions, the owner must be fairly compensated for the water.
- Private and restricted private water can be traded like any other good.
- Public water cannot be sold.
- The market sets the prices.

Most scholars agree that the government must intervene to fix prices when a merchant's conduct harms the market or the public welfare (Sabeq 1981). Muslim scholars also state that whenever the interests of the merchant and those of the consumer clash, the interests of the consumer must be given priority. Scholars agree that Islam forbids speculation and manipulation of the market to raise prices and increase profit. It is reported that the Prophet said, "whoever enters in the affairs of Muslims to manipulate prices, it is rightful for Allah to seat him in hellfire."[8]

Implementing water demand management through pricing

Islam supports a free market that is based on accessibility, fairness, and social justice. Therefore, water pricing implementation in a Muslim society is not different from elsewhere. Bhattia et al. (1995) define demand management as any measure designed to reduce the volume of fresh water being withdrawn, but without reducing consumer satisfaction or output, or both. Such measures, which are consistent with sustainability, include creating market and non-market incentives and developing institutional focus.

Market incentives

The goal of market policies is to align private incentives with social equity, thus reducing the need for co-ordination and control by governments. Price is the most direct market incentive because users alter their market behaviour in response to their private costs. Price matters in developing countries as elsewhere; price elasticities of demand are consistently found to be negative and significant, varying between −0.3 and −0.7, and averaging −0.45. This means that, everything else being equal, a 10 per cent increase in water price will lead to a 4.5 per cent reduction in demand. Despite this fact, there is still a misconception in many countries that water prices do not play a significant role in determining water demand, because water bills constitute only a small fraction of total household expenditures and total industrial production costs (Cestti et al. 1996).

Ironically, raising prices of piped water can actually benefit the poor, who pay very high prices to street vendors, provided that they can connect to the municipal system. The cost of the next water supply project can be two to three times the cost of the current project. Because prices are already subsidized, a move to full-cost pricing often will mean increasing water rates by six or seven times (Bronsro 1998). However, this still leaves room for manoeuvre if the poor are already paying five to ten times the official rate (Arlosoroff 1993).

Other market-based direct catalysts include tax incentives for investment in water-saving technologies by industries, rebates for low-water-use appliances in homes, as well as loans, discounts, and technical assistance. Finally, a market-based method of signalling the opportunity cost of water is the use of water auctions, water markets, and tradeable water rights. As of 1995, Chile was the only developing country with a comprehensive set of laws to encourage water markets (Bhattia et al. 1995). However, this market approach should not be left without control, because water is an essential commodity and poor people should be guaranteed access to basic needs. This issue is not discussed here and could be the subject of further investigation.

Institutional focus

Institutional culture can be positive or negative, enabling or obstructing. As shown in the previous sections, the problem in Muslim societies is not the lack of an appropriate culture of water demand management, it is rather the challenging task of implementing it. This in itself is an important topic that needs further study.

A concern with institutions implies acceptance of the evolutionary nature of institutional change, and acceptance of much longer time frames than international financial institutions historically have dealt with. Emphasis on institutional reform is not new in the water development field; the World Bank has promoted local reform and capacity building for at least thirty years. However, the traditional approach is characterized by impatience or, to use Thomas Callaghy's (1994) term, analytic hurry. It also tends to take institutions as given and fixed, and as all-powerful enforcers of obligations and guarantors of rights. Adding institutional elements to the traditional economic viewpoint can address such issues by merging theory with economic history, as suggested by Myrdal (1978) and others. Callaghy (1994) stresses that aid agencies must accept that change in developing countries occurs slowly and unevenly, and depends upon complex factors. The hard work of implementation is still to come. The success of water pricing as a water demand management initiative will depend on promoting "a new cultural appreciation that water is a limited resource for which people of the area must pay" (NRC 1995).

Although Islam puts forward a coherent set of guidelines and principles for a fair and effective management of water resources, numerous Muslim countries have been experiencing market failures, obstacles to new ideas, absence of institutional focus, and unfair water distribution practices. Implementation of Islamic principles must go through a stepwise and lengthy process of change.

Notes

1. 2:245.
2. Al-Muwatta 36.27.
3. Abu-Dawood 3444.
4. 59:28.
5. Abu-Dawood 3470.
6. Ahmad 524, in *Hadith Encyclopedia*.
7. Al-Bukhari 2.549.
8. Ahmad 19426, in *Hadith Encyclopedia*.

REFERENCES

Arlosoroff, S. (1993), "Water Demand Management in Global Context: A Review from the World Bank," in D. Shrubsole and D. Tate (eds.), *Every Drop Counts: Proceedings of Canada's First National Conference and Trade Show on Water Conservation, Winnipeg, Manitoba*, Canadian Water Resources Association, Cambridge, Ont.

Bahl, R. W. and Linn, J. F. (1992), *Urban Public Finance in Developing Countries*, Oxford University Press, New York.

Bhattia, R. and Falkenmark, M. (1993), *Water Resources Policies and the Urban Poor: Innovative Approaches and Policy Imperatives*, World Bank, Washington, D.C.

Bhattia, R., Cestti, R., and Winpenny, J. (1995), *Water Conservation and Reallocation: Best Practice Cases in Improving Economic Efficiency and Environmental Quality*, World Bank, Washington, D.C.

Bronsro, A. (1998), "Pricing Urban Water As a Scarce Resource: Lessons from Cities around the World," in *Proceedings of the CWRA Annual Conference, Victoria, B.C., Canada*, Canadian Water Resources Association, Cambridge, Ont.

Callaghy, T. M. (1994), "State, Choice and Context: Comparative Reflections on Reform and Intractability," in D. E. Apter and C. C. Rosberg (eds.), *Political Development and the New Realism in Sub-Saharan Africa*, University of Virginia Press, Charlottesville.

Cestti, R., Guillermo, Y., and Augusta, D. (1996), *Managing Water Demand by Urban Water Utilities*, World Bank, Washington, D.C.

Crane, R. (1994), "Water Markets, Market Reform and the Urban Poor: Results from Jakarta, Indonesia," *World Development* 22 (1), pp. 71–83.

Hyden, G. (1983), *No Shortcuts to Progress*, University of California Press, Berkeley.

Lovei, L. and Whittington, D. (1993), "Rent Extracting Behavior by Multiple Agents in the Provision of Municipal Water Supply: A Study of Jakarta, Indonesia," *Water Resources Research* 29 (7), pp. 1965–74.

Myrdal, G. (1978), "Institutional economics," *Journal of Economics Issues* 21, pp. 1001–38.

NRC (National Research Council) (1995), *Mexico's City Water Supply: The Outlook for Sustainability*, National Academy Press, Washington, D.C.

Panayotou, T. (1993), *Green Markets: The Economics of Sustainable Development*, ICS Press, San Francisco.

Sabeq, S. (1981), *Fiqh essounna* [Understanding the Prophet's tradition] (3d ed.), Dar El-Fiqr, Beirut.

Serage El-Din, I. (1994), *Water Supply, Sanitation, and Environmental Sustainability: The Financing Challenge*, World Bank, Washington, D.C.

World Bank (1992), *World Development Report, 1992: Development and the Environment*, World Bank, Washington, D.C.

——— (1993), *Water Resources Management*, Policy Paper, World Bank, Washington, D.C.

Zouhaili, O. (1992), *Al-Fiqh wa-dalalatuh* [Islamic jurisprudence and its proof], Dar El-Machariq, Damascus.

9

Ownership and transfer of water and land in Islam

Dante A. Caponera

Because Islam arose and developed in a desert area where water resources were extremely important, Muslim sources and scholars have much to say about the ownership and transfer of water and of land tenure. The environment, however, was not the only reason for this. It was also connected with the nature of Islam as a monotheistic religion that sought to regulate the behaviour of humans according to the commands of Allah.

Before the Prophet Muhammad, in the *djahilyya* or "period of ignorance," water regulations were not established in Arabia. Wells belonged either to an entire tribe or to an individual whose ancestors had dug it. In either case, the tribe or the individual proprietor of the well charged a fee to all strange tribes who came to draw water for themselves or their animals (Caponera 1973). In the south of Arabia where water was plentiful, ownership was individual and was even divided up into infinitesimal allotments. Selling water was a common practice. In general, however, water was scarce for the settled populations and the nomads, and its possession was the object of many bloody struggles: force made the law.

The Prophet Muhammad, on the other hand, preached charity as the principal virtue, inasmuch as it involved helping the unfortunate and showing detachment from material things. Thus, starting from this general principle and according to the word of Allah that "*Then shall anyone who has done an atom's weight of good, see it! And anyone who has done an atom's weight of evil shall see it*,"[1] the sharing of water appeared to the

Prophet to be an act of religious charity, and subsequently became in most cases a legal obligation. The Prophet also declared that access to water was the right of the Muslim community – no Muslim should want for water – and the Holy Quran has sanctioned this with the general formula: "*We made from water every living thing.*"[2]

Furthermore, the Prophet Muhammad declared that "Muslims have common share in three (things): grass, water and fire"[3] and, to prevent any attempt to appropriate water, he prohibited the selling of it (Yahya ibn Adam 1896, 75): the specific *hadith* related to this states, "Allah's Messenger forbade the sale of excess water."[4] On the basis of these later *hadiths*, some authors came to believe that the Prophet had established a community of water use among men (Van Den Berg 1896, 123).

It was to prevent water from being seized and hoarded by one person that the Prophet endeavoured to ensure that all members of the community had access to water. On his advice, Othman bought the well of Ruma and made it into a *waqf* (a usufruct or collective property for religious purposes and public utility) for the benefit of the Muslim community).[5] He also proclaimed that high-lying areas should be irrigated before low-lying areas; and to prevent the hoarding of water, he ordained that the quantity of water retained should not reach over the ankles.[6] In addition, the Prophet recognized that the ownership of canals, wells, and other water sources entailed the ownership of a certain area of bordering land or *harim* on which it was forbidden to dig new wells so as not to damage the quality or lower the quantity of the water in the existing ones (Yahya ibn Adam 1896, 75).

Besides these fundamental revelations which are universally recognized by Muslims of all rites, sects, and schools, other principles are found in later *hadiths*, the genuineness, or at least the interpretations of which have been contested. Scholars of the two major branches of Islam, the Sunnites and Shi'ites, by interpreting the inner meaning of the Prophet Muhammad's prophecies, sought to adapt the principles to local exigencies arising from more complex situations – in particular issues relating to the right of thirst, to irrigation, and to the sale and transfer of water and land.

The right of thirst

The right of thirst is juridically the right to take water to quench one's thirst or to water one's animals. This right is recognized to both Muslims and non-Muslims.

According to the Sunnites, the right of thirst applies to water everywhere (Al-Wanscharisi 1909, 283).[7] This principle, however, may be considered as being one of public utility, depending on the category to which these waters belong. The three main categories of water (private goods, restricted public goods, and public goods) in Sunnite doctrine are outlined in the chapter by Kadouri, Djebbar, and Nehdi in this volume. In Shi'ite doctrine, on the other hand, the right of thirst is limited to public waters (unowned waters, sources, and wells). In the case of privately owned waters, no one other than the proprietor is entitled to their use, and whoever takes of this water must return an equivalent amount (Querry 1872, arts. 69–73).

Irrigation

In Sunnite doctrine, community rights apply only to large bodies of water (Ali ibn Muhammad 1903–8, 313). A distinction must be made between lakewater, which can be used for all irrigation purposes without any objection; riverwater, which can be used for irrigation provided that it does not harm the community; and rainwater which, falling on land without an owner, is at the disposal of anyone for irrigation. The owner of the nearest cultivated plot has first priority. If there are several cultivated plots near the water, no order of priority is observed; however, the owner whose crops are most urgently in need of water takes first turn (Ahmad ibn Husain 1859, 900; Khalil ibn Ishak 1878, secs. 16.1, 20.2, 20.3).

Irrigation rights of private individuals may involve acts of individual appropriation, and in Sunnite jurisprudence are subject to different rules depending whether the rights are over small rivers where the water must be stored to raise it to the required level, canals, wells, or springs and rainwater.

For small rivers where the water must be stored to raise it to the required level (Ali ibn Muhammad 1903–8, 313 and 322), two general principles govern irrigation rights. When water is scarce, upriver pieces of land are irrigated first, but the quantity of water retained should not reach above the ankles; otherwise one can irrigate as much as one likes (Khalil ibn Ishak 1878, secs. 19–21).

Concerning the quantity of water that the owner of an upriver plot should return to a downriver plot for irrigation, the *Shafi'i* consider that only the surplus water (that which remains standing in his fields after the ground is saturated) should be returned, but the *Maliki* hold that an upstream owner should not artificially hold back any water after he has irrigated his land, but should allow the remainder to flow back to lower-lying lands without waiting for the water to completely saturate his fields. If as a result the lower-lying plots are inundated, he is not required to pay damages, provided that it was not done out of spite or carelessness (Ali ibn Muhammad 1903–8, 315).

Irrigation canals are the joint property of the individuals who built them, and they alone are entitled to exercise the right of irrigation (Ali ibn Muhammad 1903–8, 316: Al-Wanscharisi 1908–9, 285). For other construction works (mills, bridges, and so on), the consent of all co-owners is required (Ali ibn Muhammad 1903–8, 316: Al-Wanscharisi 1909, 285). The manner of use should be established by mutual agreement among all involved (Ibn 'Abidin 1869, 439).

The digger of a well, whether on his own land or on unoccupied land, becomes the owner of the wellwater as soon as he has finished digging it (Ali ibn Muhammad 1903–8, 321). Possession through use is also a subject of discussion (Muhammad ibn Ali 1923, 169). The owner of the well is the sole holder of the right of irrigation and is not required to supply water to irrigate other land (Ahmad ibn Husain 1859, 90–91; Khalil ibn Ishak 1878, secs. 18, 19; Ali ibn Muhammad 1903–8, 319–20).

The *Maliki* stress that a gift of surplus water to an owner whose well has caved in through no fault of his own is obligatory and is made without payment; however, if the cave-in is due to his carelessness, he may have such water only if he pays for it (Khalil ibn Ishak 1878, secs. 18, 19; Malik ben Anas 1911, 190–91). The *Shafi'i* consider that it is always obligatory to give one's surplus water for the irrigation of the fields of others. The *Hanifi* say that there is never any obligation incumbent upon the water owner.

Anyone who digs out or improves a spring in unoccupied land has the exclusive right to irrigation (Ali ibn Muhammad 1903–8, 321) and rainwater belongs to the owner of the land on which it falls (Khalil ibn Ishak 1878, secs. 16.1, 20.1). On no account, however, can surplus springwater and rainwater be refused for the irrigation of land where crops are in danger of dying.

The general Shi'ite principle of irrigation rights is that these belong solely to the title holder of the source of water in question, free of any servitude. Where there are several owners, the distribution of water among them depends on whether the source consists of springs, wells, or rain; an artificial canal; or a natural watercourse.

When the water supply from springs, wells, and rainwater is sufficient to supply everyone's requirements or when the proprietors agree on the manner of possession, no difficulties exist. In contrary cases, however, the water is divided proportionately to the size of the respective plots, with due consideration to the location of the land (Querry 72, art. 74). The waters of an artificial canal, on the other hand, become the property of the diggers, and the right of irrigation is exercised in proportion to the funds invested (ibid., art. 75). In the case of natural watercourses, upstream landowners are entitled to first use of the water – for crops, the plants should be covered with water; for trees, the foot of the tree should be under water; and for date palms, the water should come to trunk height. The upstream proprietor is not obliged to let the water reach the plots downstream until he has finished irrigating his own crops in the described manner, even if the crops of downstream owners suffer as a result (ibid., arts. 76, 77).

Transfer and sale of water ownership

In Sunnite jurisprudence, the *Maliki* and *Shafi'i* follow the principle that the owner of a supply of water may sell and dispose of it at will, except in the case of water in a well dug for the watering of livestock (Khalil ibn Ishak 1878, art. 1220, secs. 16, 17; Malik ben Anas 1911, 122; Ali ibn Muhammad 1903–8, 320). The purpose of the sale must, however, be known and stipulated. The *Hanifi* and *Hanbali*, on the other hand, only allow the sale of water in receptacles (Ibn 'Abidin 1869, 441).

By contrast, the right of irrigation is attached to land and follows it in all transactions involving the land. Though the owner may dispose of the land without the irrigation right, doctrines differ as to the right of disposing of the irrigation right in this case. The *Hanifi* do not permit the sale of the irrigation right, which can only be transferred by inheritance. However, the owner can attach the irrigation right to another piece of land without such a right that he owns or of which he acquires ownership – and the irrigation right can then be sold with the land, thereby enhancing its value (Ibn 'Abidin 1869, 441). The *Maliki*, on the contrary, allow full freedom of action in regard to the disposal of the irrigation right. In particular, they recognize the right to sell it, reserving the use of the water to certain specific days. They also recognize the right to sell shares of irrigation time while retaining possession of the right itself and its sale or rental apart from the land (Malik ben Anas 1911, 10:121–22).

Under Shi'ite principles, on the other hand, water can be sold only by weight or by measure, i.e., it must be in a container, because of "the impossibility of delivering it owing to the possible immixture of extraneous substances" (Querry 1872, art. 67).

Land tenure and water rights

Islam began with no administrative machinery; it therefore developed on a customary basis. Land ownership as it exists in Islam was mostly determined by Muslim land laws, which developed during the centuries following the Muslim conquest, largely on the basis of the Byzantine concept of supreme ownership by the ruler of the state.

Land taxation practices developed according to the general examples given by the Prophet. The population was divided into two categories: Muslims and non-Muslims or *dhimmis*. Muslims paid a tax called *usher* (tenth or tithe), which varied between 5 and 10 per cent of the value of the harvest according to whether the land was irrigated (either artificially or naturally) or not. *Dhimmis* payed two different forms of taxes: the *jizya* and the *kharaj*, which soon became to mean respectively "poll tax" as a tribute for protection and "land tax."

"Muslim community" is the expression that Muslim jurists use to designate the state, and "imam," originally the *khalifa* and later the sultan, to designate the qualified representative of the community. Imams, as a matter of principle, never had any legal authority or power in classical law to control the distribution of waters irrigating private land (*miri* property, with the owner having the full right of disposal). Their authority, however, does extend to water attached to *miri* property, that is, property in the collective ownership of the entire Muslim community.

The ultimate owner of *miri* property is the state, while the landowner has the status of a quasi-owner. He may sell, let, mortgage, or give away ownership, but cannot bequeath it by will. In practice, the estate can be inherited by sons although this was not allowed in the beginning, but if there are no heirs the property goes back to the state. The state has a right of supervision. The theory that land given for the purpose of cultivation must be cultivated by the recipient or occupant and that he must pay taxes is upheld. The validity of any transfer of such lands must be certified by the state or its agents.

There are many different forms of collective ownership, the most important being: *Mawat*, *mewat*, or *mushaa*; *kharaj*; and *waqf*.

Mawat, *mewat*, or *mushaa* are uncultivated "dead lands." They are considered as being in the collective ownership of the Muslim community in Arabia, Iraq, Jordan, Lebanon, and Syria. This form of land ownership allows the individual only a share in the possession of the land which is owned collectively by the village or tribe; there is no individual right of ownership. A system of rotation enables each person to receive a different share each year. Although the absolute power of the *khalifa* to make land grants out of such idle tracts is recognized, either by granting ownership of both the soil and the waters thereon or by allocating titles to water and land separately, other concepts have been developed by the various schools of law. The *Hanifi* claim that there cannot be private appropriation of land without cultivation, even with the permission of the sovereign, and the *Maliki* claim that land can be owned privately with such permission provided it is developed (Malik ben Anas 1911, 15:195).

Kharaj or conquered lands are cultivated and productive lands on which the *kharaj* or land tax is levied, as is done on all conquered lands from which the sovereign has neither expelled nor expropriated the inhabitants, whether or not they have converted to Islam. Being the property of the Muslim community, these lands are administered by the *khalifa*. The owner, in principle, does not hold full title to the property but only enjoys the usufruct from it. Muslim administrative authorities were responsible for all questions dealing with waters on these lands.

Waqf is land owned by the state, the income from which constitutes state revenues, and is allotted to pious foundations – mosques, cemeteries, fountains, schools, and so on.

Present-day practice

Water resources in Islam are public property (state property or public domain). This facilitates the proper management and administration of water. In fact, most Muslim countries that have passed recent water legislation have declared all water to be part of the state or public domain. In this way, it follows that a permit or concession is required for any use of water. In these permits, which are temporary (from one to fifty years), the water administration may insert all the conditions it considers necessary, on the basis of plans or in the public interest.

The same procedure is followed with regard to the payment of water rates, fees, or other financial requirements. If, in theory, it is not possible to tax water in itself because it is a gift from God, it is perfectly legitimate to tax the water service or to tax the supplying of water for different purposes, always with a permit. This is the practice in many Muslim countries.

The transfer of water can also be handled as the water administration wishes. It may reduce, under certain conditions, the right to use water and transfer it to another user. If all the waters are to be taken away from a group of users, always for legitimate purposes, the administration may do so in appropriate circumstances and against compensation.

Islam imposes no restrictions on trading water. Water, being a public property, cannot be transferred, but its use can. Therefore, if a user, large or small, possesses a water use permit or concession, he may trade this water to another user, large or small, if the water administration, which is the trustee for public water, so allows.

In Muslim countries, fragmented water laws and inefficient water institutions have been responsible for the mismanagement of water resources.

This is because comprehensive legislation and proper institutions to enforce the law are lacking. For example, water legislation is needed to control pollution of groundwater, particularly in shallow aquifers, caused by the discharge of untreated wastewater. Similarly, a permit system is needed to control pollution by setting maximum discharge levels and the standards to be maintained. In addition, it is most important to have a comprehensive water rights administration to control all uses of water. The Expert Group Meeting on Water Legislation of the UN Economic and Social Commission for Western Asia (ESCWA), held in Amman on 20 November 1996, concluded that the "integrated management and development of water is contingent upon the establishment of an effective legislative framework for an integrated approach to the regulation, development and management of water and other related water activities" (ESCWA 1996). The passage of water laws emphasizing the management of water resources is indeed needed in all Muslim countries, and the religious precepts of Islam are not an obstacle to the proper management of water resources in all of its aspects.

Notes

1. 99:7–8.
2. 21:30.
3. Abu-Dawood 3470.
4. Muslim 3798.
5. Al Bukhari 2.102, in *Hadith Encyclopedia.*
6. Al-Bukhari 3.550.
7. Al-Bukhari 2.104.

REFERENCES

Ahmad ibn Husain, Abu al-Shuja, al Isbahani (1859), *Précis de jurisprudence musulmane selon le rite des Chaféites*, tr. Keijzer, E. J. Brill, Leiden.

Ali ibn Muhammad, al Mawardi (1903–8), *Traité de droit public musulman*, tr. L. Oshorog, Leroux, Paris.

Al-Wanscharisi, Ahmad (1909), *La pierre de touche des Fetwas*, tr. E. Amar, vol. 2, Leroux, Paris.

Caponera, Dante A. (1973), *Water Laws in Moslem Countries*, FAO Publications 20, no. 1, Organisation. Food and Agriculture Organisation, Rome.

ESCWA (UN Economic and Social Council, Secretariat) (1996), *Water Legislation in Selected ESCWA Countries*, Publication E/ESCWA/ENR/1996/WG.ll/WP, ESCWA, Amman.

Féhliu, E. (1909), *Etude sur la législation des eaux dans la Chebka du Mzab*, Mauguin, Blinda.

Ibn 'Abidin (1869) (1296), *Al dorr al mokhtar* [The chosen jewel], vol. 5, Beulag.
Khalil ibn Ishak, al-Jundi (1878) *Code musulman par Khalil, rite Malékite*, tr. N. Seignette, A. Jourdan, Algiers.
Malik ben Anas (1911), *Le Mouwatta: Livre des ventes*, vol. 15, tr. F. Pelier, A. Jourdan, Algiers.
Muhammad ibn Ali, al Sanusi (1923), *Kitab chifa l'sadar bi arial masail al'achri* [The book of thirst by Sadar], vol. 8, Imprimerie Qaddour ben-Mourad al-Turki, Algiers.
Querry, A (1872), *Recueil des lois concernant les musulmans Schytes*, vol. 2, Imprimerie Nationale, Paris.
Van Den Berg, L. W. C. (1896), *Principes du droit musulman selon les rites d'Abou Hanifah et de Chafei*, tr. De France and Damiens, Algiers.
Yahya ibn Adam (1896), *Kitab al kharadj: Le livre de l'impôt foncier*, E. J. Brill, Leiden.

10

Water markets and pricing in Iran

Kazem Sadr

The water market has had an important role in the provision and distribution of water since the rise of an Islamic state in Arabia and has continued performing this function as the economies of Muslim countries have developed. This chapter discusses the experience of Iran with respect to the structure and behaviour of the water market and describes the innovations that have taken place in alternative forms of exchange and pricing practices before and after the Islamic Revolution.

Water resource ownership and utilization rights

Rights to ownership of water resources are discussed in the Islamic law literature, or more precisely in *fiqh*, along with property rights over mines. The latter are grouped into shallow or "open" mines and deep or "inward" ones. Water is generally considered to belong to the former group, and is discussed together with it. It is the consensus of the *fuqaha* (Muslim jurists) that both surface and underground water sources are

I am indebted to my colleague Mr A. Noori Isfandiari who encouraged me to write this essay. Ideas and information provided by him appear frequently; however, I am responsible for all errors. I am also grateful to Dr H. Ghanbari who devoted considerable time to refereeing and editing this chapter; and to the International Development Research Centre (IDRC) for providing me the opportunity to participate in the Workshop on Water Resources Management in the Islamic World.

either a common property resource (Ibn Barraj 1410 A.H., 6:257–58) or a part of *anfal* – a property of the Imam, the just and legitimate ruler, which can be operated directly by the government or leased out to private agents (Kolaini 1388 A.H., 1:538).

Investment by any share holder for the purpose of gaining access to these sources grants him private ownership of or a priority right to the use of the water that is so obtained, but gives him no claim to the river or reservoir from which the water comes. Wells, *qanats* – series of wells whose bottoms are linked by gently sloping underground canals in which water flows by gravity – or channels, which are alternative forms of investment for gaining access to water, are the private property of the investor. The water that is pumped or channelled in these ways belongs to the investor too. However, the water source as such remains the common property of the community.

While no one can "own" the water source itself, in some cases, depending on the nature of the source, one can gain exclusive water use or withdrawal rights. The different cases are outlined below.

Types of rights to water sources

To begin with, seas, lakes, and large rivers are all common property in Islamic law and no one can appropriate them exclusively. Toosi (n.d., 3:282) has declared consensus among the *fuqaha* on this point. Both Iranian civil law (article 155) and the Constitution of the Islamic Republic of Iran (article 45) make the same assertion. In any case, the water supply from these sources usually exceeds the demand, and thus no one earns exclusive or even priority rights to exploitation. Everybody has an equal right to withdraw water.

Again, if water naturally flows out of springs or through canals without anyone's effort or investment, it is similarly the common property of all. The flow of water through these sources may soon fall short of demand, however, because of population or economic growth. Therefore, an allocation criterion needs to be defined. Some *fuqaha* have offered "first come, first served" as a basis. Anyone who precedes others in obtaining access will obtain priority right to use the flow of water; the stock of the surface or underground resource remains, however, as a common property of the community.

The basis of this "preceding rule" is a *hadith* that states that anyone who precedes others in using a property deserves it most (Beihaqi n.d., 6:142; Noori 1408 A.H., 4:6). However, this priority right does not allow the user to appropriate more than he needs because the property is still commonly owned and the preceding principle does not negate the rights of others. Needless to say, the preceding privilege does not create an outright possession right.

If the supply of water through a commonly owned resource would not even satisfy the legitimate demand of all partners, how should it be distributed among them? Some *fuqaha* have suggested drawing lots. Others, however, give priority according to distance from the source, so that farms will be irrigated one after the other, and the last drops will be used by the farthest farm. Najafi (1392 A.H., 38:110) preferred this criterion over the first. This procedure is also based on a *hadith*, and it has been followed in many Muslim countries. The Iranian civil law (article 156) clearly states that if a stream of water is not sufficient to irrigate all adjacent land and a dispute arises among the landholders, none of whom can prove his priority right, then whoever is closer to the water should precede those farther away, and irrigate just as much as he needs.

In cases where access to a common pool of water is obtained by either drilling a well or making a canal, then the investor earns private property right over the water withdrawn. Najafi writes that when commonly owned water is contained (in a pool or canal), then it becomes an exclusive property of the *haez* (the one who has contained the water) provided that in so doing he does not harm others. Najafi (1392 A.H., 38:116) further adds that there is no difference of judgement among *fuqaha* in this case. Toosi (n.d., 3:282) declares if anyone steals such water, he is obliged to return it to its owner. Articles 149 and 150 of Iranian civil law recognize the same right.

Where a person digs out a well in his own farm or in arid land, intending to withdraw water, the majority of *fuqaha* believe that he will become the sole owner of both the well and the water (Najafi 1392 A.H., 38:116). However, Toosi (n.d., 3:282) declares that he only deserves a use permit and cannot sell the water that is in excess of his need. Toosi's verdict is based on a few *hadiths* in Ibn Abbas, Jaber, and Abu Horairah, quoting the Prophet (pbuh) to the effect that selling excess water is impermissible (Beihaqi n.d., 6:151). The majority of *fuqaha*, however, argue that these narrations cannot limit the right of free exchange. Not only is the latter rule general and unconstrained, but there are also other traditions that specifically permit the exchange of the extra water. Thus the quotations from the Prophet (pbuh) are assumed to mean that selling the water in excess of one's need prior to *hiazat* is not permissible, or that such a transaction is not recommended.

Both Imam Sadegh and Imam Mossa Ibn Jafar approve the sale either for cash or for wheat of one's share of a *qanat* (Al-Hurr al-Amiliyy 1403 A.H., 277–78, 332).

Thus the majority of *fuqaha* hold the view that if anyone can obtain occupancy right over a stream of water channelled or pumped from a commonly owned source, he can sell freely all or part of his lot. The same right is recognized in article 152 of the Iranian civil law.

Government and water rights resources

Water resources are common property of people and not a government domain. Thus everybody has equal right to withdrawal, and this private activity is honoured and cannot be interrupted as long as it does not harm anybody. But the exercise of this right may lead to overuse, and underground water reservoirs are particularly susceptible to exhaustion through overpumping. In such cases, the rule of "no harm" or "no overuse" will overrule freedom of the operation. Government authorities at the local or national level then will proceed according to the rules discussed earlier in pursuit of the public interest. Other rules that serve the same purpose are discussed in the next section.

Governments may also sometimes need to resolve the conflicts that may arise among competing users of common water property. For instance, construction of dams in rivers usually increases the supply of both drinking and irrigation water, and expansion of agricultural activities or rapid growth of population, or both, may lead to shortage of water for either or both sectors. In such cases, government can interfere to determine the utilization priority. This will result in depriving one beneficiary group of sufficient access to water, thus bringing it under the "no harm" rule. By compensating the losers, governments can resolve the problem.

Government and water markets

The early Islamic state

One of the characteristics of an Islamic economic system is that economic activities are totally delegated neither to market organizations nor to public sector planning boards. Economic affairs are divided between the two sectors and each carries out its own provision, allocation, and distribution function. In fact, the two prominent economic institutions at the time of the Prophet (pbuh) and his successors were the market, which carried out the supply and distribution of private goods, including water; and the public treasury or *baitulmal*, which was responsible for economic planning and establishment and operation of infrastructure investments, including dam reservoir construction.

During the early Islamic era, many participants were active in each market, and their behaviour was controlled by inspectors (Sadr 1996).

Sellers and buyers could freely enter or leave any market to choose the best enterprise based on available information. The government's right to interfere in the market to set prices was limited. On the basis of this early practice, a general agreement – although not a consensus among the jurists – seems to have arisen that if the market is behaving well, then nobody is permitted to interfere by setting prices. On the other hand, governments may do so if prices are fluctuating and equilibrium cannot be restored in the market government (Rajaee 1996, 57–98). Regarding the criteria for price setting, most *fuqaha* insist on a "just price," a price that will be determined in the market if the rules of *sharia* are operative and the market is in the normal condition (Khomeini 1989, 4:318–19). Otherwise, the price to be set must be equal to the equilibrium price under normal conditions. This criterion is usually called "the likeness value" in the *fiqh* literature (Toosi 1404 A.H., 4:23).

Early Islam likewise set precedents for prevention of hoarding, waste of commodities or inputs, and imposition of external costs on neighbouring operators, which, in addition to full observation of Islamic codes of contract, contributed to an efficient exchange in the market. The absence of quotas, customs, or tariffs further facilitated trade. Thus the prices that were determined in the market were *efficient*, that is, no other prices, if set, could further increase consumers' satisfaction or sellers' profits (Sadr 1996, 188).

The rise of the water market

In many parts of the world, such as Africa and Asia, water has been a cause of human settlement and civilization (Issawi 1971, 213). People have settled around rivers and springs to be able to make a living in dry climates. In the initial stages of development of such communities, water supply usually exceeds demand. However, at later stages of growth, because of the rise of population, income, and a variety of economic activities, the demand for water increases and eventually exceeds the supply, and water usually comes to be rationed through the guidance of the community's norms and customs. Because the rationing methods are suggested by the community members themselves, they are coherent and in conformity with the community's accepted set of rules and rights, and lead to legitimate devices for water distribution.

Over time, in growing human societies when demand exceeds supply, new market institutions are created, because the existing sets of rules and traditions fall short of an efficient allocation. In such segmented water markets, whose size depends on the supply of water, the most reliable and accessible means of exchange is water itself, because it can be used to produce any crop. In some parts of the Middle East, for example Iran where 80 per cent of the land under cultivation is used for wheat and barley, it is natural that these crops would serve as means of exchange in the market for water. This phenomenon, that is, transactions in kind rather than in cash, may have caused the impression that water has not been a commodity and has not been sold or bought in the market.

The legal system of rights in Islam, as mentioned earlier, recognizes the market institution for water transactions. The cases that have been reported by Safinejad (1985, 1996) and other anthropologists present the evidence. The media of exchange, in their reports, are the staples – food and water – and seldom money.

Private and public water supply

The market is not the only institution that manages the supply and demand of goods and services in communities. Many public firms and group organizations have been formed to carry on the same function. Buchanan (1968) foresees variable but continuous types of organizations supplying or allocating public and private goods. His analysis is based upon the external decision-making cost of providing these commodities (Buchanan and Tullock 1971). He anticipates the formation of markets for private goods, group or communal organizations for public goods, and government takeover of the pure public goods (Buchanan 1968). His prediction has been validated in many societies, but in Muslim countries, water is a commodity for which all three types of organization have been used, because it can be considered as a private good at some times and a public one at others.

Monopoly and government supervision

In many economies, the market for gas, water, electricity, and telephone service tends toward a natural monopoly structure. The share of initial fixed investment for provision of these services is high and that of variable costs is low. As a result, the average variable and marginal cost of offering a new extension or serving a new customer is very low; no other supplier can compete with the one who is already in the market. This monopoly situation and the high cost of arbitrage among consumers tempt sellers to indulge in price discrimination. Thus water is sold at different prices to urban consumers, industrialists, and farmers.

Another type of discrimination is to reduce prices as the quantity purchased increases and so to encourage the customer to buy more. Recently, realizing that the demand for water is inelastic, many sellers have followed an increasing-block pricing scheme (Sadr 1996). Finally, sellers are able, at times, to practice perfect discrimination using both techniques together. These practices lead governments to supervise the performance and price strategies of the utilities.

Water pricing practices in Iran

In Iran, major rivers flow mainly in mountainous areas where surface water is the main source of irrigation. The rest of the country depends on underground water drawn up from *qanats*.

Surface water

Rivers are used by farm operators on the basis of proximity (article 156 of the civil law). As Lampton (1969) reports, the village of Toroq, near Meshed in northeastern Iran, is irrigated after the villages that lie closer to the local river. The same is the case in Kurdestan, where the villages located close to rivers use as much water as they need, and whatever is left is allocated to the distant villages. However, no one can build a dam or a floodgate in the fields through which water flows. During the summer season, the flow of water in many rivers declines so that villages that have appropriated water rights have priority to use the water. For example, Lampton reports that the water of Zayanderood is distributed according to practice dating from Safavid times. From 15 November until 5 June, withdrawal of water is unlimited. However, in the summer, water is allocated to certain regions and villages. From Jadjrood, river water is also distributed according to an old tradition: some counties have appropriated rights, and others must pay for it.

Since 1943, provision and management of surface water has been officially controlled by a governmental water agency (Ministry of Energy 1994, 16–21). Later, regional water organizations were established that monitored dams in each region and distributed water among villages. As of 1968, after the enactment of the Nationalization of Water Law, regional water agencies were required to charge for the distributed water enough to cover their average expenses, which should include variable costs of maintenance and the fixed cost of depreciation and interest (Ministry of Energy 1994, 392). In 1982, the same law was revised, extended, and approved by the parliament as the Just Distribution of Water Law. The irrigation water must be priced on the basis of average variable cost and depreciation as before, but interest is not included. In areas where metering is difficult, water can be charged for by the farm size and the type of crop (Ministry of Energy 1994, 234–40).

The procedure that is approved by the Ministry of Energy for agricultural water charges as of 1990 is as follows.

- The average price of water withdrawn from "modern networks" – that is, primary and secondary canals of dams – is 3 per cent of the total revenue of the crops planted; of water from traditional canals, 1 per cent; and of water transferred by a combination of the two, 2 per cent.
- The average production of crops in each region is obtained from the yearly statistics of the Ministry of Agriculture. The value of each crop is measured either by the guaranteed price, if there is one, or by the farm-gate price. Using these data, water agencies determine the price of water per cubic metre (Ministry of Energy 1994, 295–96).

In 1990, municipal water and sewer companies were established after approval of the corresponding bill by the parliament. The bill states that the private sector, the banks, and the municipalities can participate in the investment and management of these plants, which will be operated as companies and according to the Trade Law. This bill, which clearly lays down the legal foundation for private sector participation in urban water affairs, indicates a policy change, too. The subscription rate for water and sewer service will be calculated and proposed by the companies' boards of trustees based upon operation and depreciation costs and go into effect upon approval by the government's Economic Council.

Consumption of water up to five cubic metres per month is exempt from any charges – to ensure that low-income families have access to water for drinking, health care, and religious obligations. An increasing-block rate is charged for higher-level consumption: that for Tehran Province in 1995 is shown in table 1. Similar rates are charged in other provinces. In 1996, the charges were increased. Monthly consumption up to 5 m^3 was exempt, and water use up to 25 m^3 was charged at the 1995 rate. However, the rate for the block between 25 m^3 and 45 m^3 was increased by 25 per cent, and the rate for water use above 46 m^3 was increased by 30 per cent. In 1998, the tariff for commercial and industrial use of water was set higher than for residential consumption. This reversed the policy of previous years.

Underground water

Qanats have been the main technology for withdrawal of water from underground reservoirs, although pumped wells have recently begun to replace them. It is natural that in arid regions of Iran, water utilization rights, types of exchanges, and pricing practices are all associated with *qanats*. Therefore, the discussion here concentrates on water markets based on this type of water withdrawal.

The water of each *qanat* is initially divided among the share holders; thus a water rotation period is observed wherever this technology is used. This period is naturally shorter in spring and summer than in other sea-

Table 1 **Increasing-block charges in Tehran Province, 1994**
(in rials[a] per cubic metre; consumption blocks are in cubic metres)

5–10	11–15	16–20	21–30	31–40	41–50	51–60	61–70	70+
15	25	30	36	67	100	133	168	300

Source: Ministry of Energy, Office of Urban Water and Sewers.
[a] US$1 = 4,000 rials in 1997.

sons because of higher evapo-transpiration and consumption use of crops. Division of *qanat* water among one or more villages that are quite far from each other (Yazdani 1985) has necessitated, over time, training skilled technicians for both maintaining *qanats* in operation and distributing the water among many farmers without loss. Thus the market for two types of jobs has arisen. The first is highly technical, requiring knowledge of building and dredging *qanats*. The second requires a talent for adopting a distribution scheme by which water loss would be minimum. In addition, the distributor must be trusted by all, because he can manipulate everybody's lot. The value put on this job has resulted in alternative forms of group decision-making to select the distributors. The common feature in all forms is that an irrigation task force is chosen by the holders of the water rights, to nominate a head distributor; the holders of the water rights must then approve the choice by majority vote (Safinejad 1985).

The dredger's wage is generally paid in kind, most frequently in the form of a share of water. In a village in Gonabad, in the northeast, any particular field is irrigated every fourteen days in summer and every twenty-one days in other seasons. Payment for dredging is made by adding one share, or day, before a field is irrigated, and "paying" this extra day to the dredger. In another village in Gonabad, the period between irrigations is increased from sixteen to seventeen days and in Ghaylen from seventeen to eighteen days and once again, the additional day of water is paid for dredging (Yazdani 1985). In a village in Yazd Province (central Iran), the four-member group of distributors were paid a wage that was equal to 18.5 hours of water that they could use on their farms or sell (Safinejad 1996). Once, in a village of Tafresh, a *qanat* was ruined by a major flood and its repair was very costly for the poor peasant owners. The landlord proposed a deal: he would pay the reconstruction cost in return for one day of water in each rotation, that is, by extending the rotation from eight to nine days (Safinejad 1985).

Over time, however, payment in cash as well as in kind has become customary. In a village of Ferdous as in other parts of the country, water was distributed by a "water clock" – a device for measuring water use, in this case in the local unit, the *fenjan*. In 1971, for each *fenjan* of water, fifty rials (US$.0125 at 1997 rates) were collected for both dredging and distribution. The same price was charged in another village in 1976, and in a rural community of Yazd in 1978, every *joraeh* of water cost one thousand rials and, in total, water share holders paid 2.6 million rials (US$650) (Safinejad 1996).

As described previously, in the initial stages of community development, the water distribution task is handled by existing norms and traditions. Eventually, as the market organization is formed, transactions will be in kind initially and in cash after the community undergoes the final stages of development. Today, in rural communities of Iran, the cash valuation of water is so common that the Statistics Department of the Agricultural Ministry can easily collect water price information in different parts of the country. This information is used to calculate the average cost of crop production and suggest a guaranteed price for wheat and other supported crops to the government.

As described earlier, the private sector is quite active in extracting water from underground resources. At present, wells are replacing *qanats* because the cost and time for construction are lower than for *qanats*. However, this advantage has caused excavation of too many wells and, hence, overpumping of water. Many underground aquifers are now under stress and further well drilling is forbidden.

The Just Water Distribution Law authorizes the Ministry of Energy to supervise withdrawal activities from underground pools. A supervision charge can be levied based on a percentage of the crop price (table 2). The charges are calculated for each region and their equivalent cash value is collected. This procedure further validates our hypothesis that, as the economy grows, markets for water will be established. Initially, the unit of value is the staple food or the water itself because this medium of exchange can accelerate transactions more than others. Eventually a

Table 2 **Percentages of crop prices authorized to be charged by the Ministry of Energy for water supervision**

Wheat	0.25
Rice	0.6
Oranges, dates, and vegetables	0.85
Pistachios and almonds	1.0
Fruit trees	0.8
Other	0.5

Source: Ministry of Energy, Office of Water Affairs.

monetary *numéraire* will be adopted as trade expands in the economy. The water market seems to have undergone such development in Iran.

Conclusions

Despite the fact that water is a pious commodity in Islamic culture and its natural sources are owned in common under Islamic law, the market has played an important role in demand and supply management of water since the rise of the Islamic state in Arabia. The system of property rights in Islam allows those who spend effort and expense to withdraw water from a commonly owned source to secure private possession rights, provided that the rights of other users are preserved. This recognition provides the opportunity for the exchange of water with other goods, that is, formation of a water market, whose various organizational forms have been observed throughout the Muslim countries. But in the early Islamic state, dam construction and water reservoir developments were financed by the *baitulmal*. These two institutions – private and public – initiated and directed water supply, transfer, and distribution activities.

Utilities tend toward a monopoly structure if both provision and distribution of service are entrusted to the market. Neither Islamic jurisprudence nor economic logic justifies privatization of the whole water sector. Instead, what is recommended here is co-ordination between the public and private sectors for handling water-related activities. Overhead investments for the provision and preservation of water will be carried out by the public sector but transfer and distribution of water will be carried out by the private sector. If Islamic rules and values prevail in the market, the price that will be determined can be expected to be efficient. This price will serve as a norm for the water that is provided and sold by the public sector, and the latter's price should cover the average total cost of operation. No discrimination in water pricing should be used in practice. This proposal is consistent with the legal system of Islam and management of water supply and demand in Iran.

REFERENCES

Beihaqi, Ahmad Ibn Hussain (n.d.), *Assonan-ul-kobra* [The great (prophetic) traditions], Daral Maarefa, Beirut.

Buchanan, J. (1968), *The Demand and Supply of Public Goods*, Rand McNally, Chicago.

Buchanan, J. and Tullock, G. (1971), *The Calculus of Consent*, University of Michigan Press, Ann Arbor.

Al-Hurr al-Amiliyy (1403 A.H.), *Wasaelueshiah* [Methods of the Shi'a], Ehia Attorath-ul-Arabi, Beirut.
Ibn Barraj, Saad-ud-Deen (1410 A.H.), *Jawaher-u-fegh* [The Jewel of the *fiqh*], Addar-ul-Islami, Qum.
Issawi, C. (ed.) (1971), *The economic history of Iran: 1800–1914*, University of Chicago Press, Chicago.
Khomeini, Roohulla (1989), *Ketabul beia* [The book of choosing a successor], Ismaeilian, Qum.
Kolaini, Mohammad (1388 A.H.), *Alkafi* [The sufficer], Darul Ketab Al Islamiah, Tehran.
Lampton, Ann (1969), *Landlord and Peasant in Persia*, Oxford University Press, London.
Ministry of Energy (1994), *Water and Electricity Legislations: From the Beginning up to 1993*, vol. 1, Ministry of Energy, Tehran.
Najafi, Mohammad Hasan (1392 A.H.), *Jawaher-ul-kalam* [The jewels of speech], Dar-ul-Kotobel-Islamia, Tehran.
Noori, Mirza Hasan (1408 A.H.), *Mostadrak-ul-wasael* [The ways of understanding], Alul Beit, Beirut.
Rajaee, Kazem (1996), "Ghaymat gozari" [Price setting in Islamic economics], M.S. thesis, Mofeed University, Qum.
Sadr, S. Kazem (1996), "Water Price Setting: The Efficiency and Equity Considerations," *Water and Development* 4 (3), pp. 44–53.
Safinejad, Javad (1985), *A Study of the Economic and Social Effects of Changing Water Rotation Period*, International Seminar on Geography, Islamic Research Foundation, Mashhad, Iran.
——— (1996), "Financing the Traditional Farm Irrigation by Qanats," *Water and Development* 4 (3), pp. 98–110.
Toosi, Mohammad (1404 A.H.), *Attebyan fee tafseer-el-Quran* [Clarity in the interpretation of the Quran], Dar Ehia Attorath-ul-Arabi, Beirut.
——— (n.d.), *Al mabsout fee feqeh-el-imamiah* [A detailed account of the jurisprudence of the Imams], vol. 3, Maktabat-ul-mortadawi, Tehran.
Yazdani, Lotfollah (1985), *The Characteristics of the Southern Khorasan Qanats and Their Water Distribution*, International Seminar on Geography, Islamic Research Foundation, Mashhad, Iran.

11

Intersectoral water markets in the Middle East and North Africa

Naser I. Faruqui

As the introduction to this volume discusses, water is rapidly becoming the key development issue in the Middle East and North Africa (MENA). The natural aridity of much of the region coupled with high population growth and urbanization is creating severe inequities. Because the urban growth rate of less-developed Muslim countries (LDMCs) in the MENA is higher than the overall average for all less-developed countries (LDCs) – 3.2 per cent versus 2.9 per cent for 1995–2015 – informal settlements in cities all over the region are burgeoning. The urban or peri-urban communities are rarely served by public utilities, either because they were unplanned or because of legal or political restrictions imposed on the utilities.

Many of the community residents rely on informal supplies of water sold by private vendors. For LDCs, on average, these families pay ten to twenty times more per unit than residents receiving piped water service – up to one hundred times in some municipalities (Bhattia and Falkenmark 1993). A literature search for prices paid by the unserved urban poor in Muslim countries revealed almost no data available on the topic. However, during the exceptionally warm summer of 1998, in Jordan, the city of Amman suffered a severe water shortage, exacerbated by an odour problem. The public was forced to buy water from vendors, and the black-market price of water delivered by truck tankers reached US$14 per cubic metre (Bino and Al-Beiruti 1998). Even under normal weather conditions, some of the poor pay a very high price in Jordan. An informal survey (conducted during an International Development Research Centre [IDRC] trip to Amman in December 1998) in the Al Hussein refugee camp in Amman found that residents who were not connected to the municipal water system were buying water from their connected neighbours for prices ranging up to US$2 per cubic metre – four times the rate paid by the served customers, whose tariff includes sanitation. This is higher than the maximum theoretical cost of US$1.80 per cubic metre for desalinizing seawater and distributing it in Saudi Arabia, an adjoining country (Abderrahman, this volume). Likewise, an IDRC-supported urban water evaluation in Jakarta found that in some cases, unserved residents spend up to 25 per cent of their income on water.

The topic of water and equity in MENA countries requires more investigation through methodical, formal studies – it is precisely because the unserved poor live in informal, often unpleasant, forgotten settlements that they are ignored by mainstream researchers. However, there is no reason to believe that the prices paid by the unserved urban poor are any less in MENA countries than in those countries for which information is available. Clearly the current situation is inequitable, and the primary water use right in Islam, that of quenching thirst (*haq al shafa*), is being compromised.

How can this situation be addressed? To optimize the available water within urban areas, municipalities will have to undertake a range of demand management options, including raising tariffs. However, although some water can be saved through domestic conservation practices, the amount is limited because people in the MENA already use water very carefully. Rapidly growing populations mean that that more water will have to be allocated for domestic purposes. For instance, Israel's policy is that as urban populations grow, the first priority in water allocation will always be for domestic–urban uses, followed by industrial needs and finally by agriculture. Eventually, given the current rate of urbanization, and an unchanging combined industrial-domestic water consumption rate of 342 litres per capita per day (LPCD), by 2030, 80 per cent of Israeli freshwater will be used in cities and industry and 20 per cent in agriculture (Lundqvist and Gleick 1997). Where will water on this scale come from? Although the ratio varies from country to country, typically water is allocated in the MENA as 10 per cent to industry, 10 per cent to the domestic sector, and 80 per cent to agriculture. Domestic demand is growing and, even with recycling, so will the demands of industry as MENA countries begin to industrialize. Therefore, the water will have to come from agriculture. What will be the mechanism of this intersectoral transfer? Many recommend allowing the market to reallocate the water. Even with low tariffs, the value of water is at least ten times higher in urban areas than it is in agriculture (Gibbons, cited in Bhattia et al. 1995, 6).

Regulated water markets have been successful in developed countries such as Chile and the United States. In 1991, during a drought period, the California Water Bank purchased water from farmers for about US$0.10 per cubic metre, representing 25 per cent more profit than they could get by planting crops. The water was then sold at an average price of US$0.14 per cubic metre, to supply critical urban and agricultural uses (Bhattia and Falkenmark, 1993). Chile's Water Law also allows transfers. The city of La Serena met its growing water demand by buying water from farmers at a much lower cost than the alternative of contributing to the construction of the proposed Puclara dam, which has now been postponed indefinitely (Postel 1995). In Jordan, the government paid farmers US$120 per hectare for not planting vegetables and annual crops in 1991, a clear case of trading established water rights (Shatanawi and Al-Jayyousi 1995).

However, can intersectoral water reallocation through markets be carried out in equitable, sustainable, and economically feasible ways across the MENA region and be compatible with Islam – a sociocultural force that shapes belief and policy in the region? This chapter examines these issues by discussing the permissibility of markets in Islam, the prerequisites required for sustainable water markets, problems associated with unregulated markets, the issue of food security, and the need to take an integrated water management approach.

Water markets in Islam

There is no point to examining the feasibility of water markets as a tool for more equitable water management in the MENA, if they are incompatible with Islam. Admittedly, Islam's impact varies by country. For example, some countries, such as Tunisia, have become quite Westernized, whereas others, such as Iran and Saudi Arabia, have constitutions based upon *sharia* (Islamic jurisprudence). But in general, Islam exerts a great influence over the three hundred million Muslims of the region.

The permissibility of water markets in Islam depends on whether the following prerequisites of water markets are religiously acceptable: that individuals or groups hold specific rights to water; that they may transfer those rights; and that they may recover the costs of trading their water rights to others.

The categorization of water in Islam into private goods, restricted private goods, or public goods is discussed by Kadouri, Djebbar, and Nehdi in this volume. As noted in that chapter, private goods and restricted private goods can be owned and sold; and if water can be priced to recover costs, and sold, it is evident that it can be traded either within or across sectors. Within sectors, especially agriculture, water has been sold in Muslim countries such as Iran – both historically, in Persia after the advent of Islam, and recently, in Iran after the Islamic Revolution. The legal system in Islam recognizes the market institution for water transactions. In both Iran and Saudi Arabia, charging a tariff to recover the costs of providing potable water is not only permissible, it is recognized in the law.

It is clear, then, that Islam allows for both private and public water markets, and the charging of tariffs to recover costs for most categories of water. The question then becomes, are intersectoral water markets to reallocate water desirable in an Islamic context?

This question can be answered by examining Islamic water law in terms of utilization priority. Islamic scholars accept that water in Islamic societies was traditionally prioritized as follows: first, for domestic purposes (the right to quench thirst and the right and requirement to be clean); second, for domestic animal watering; and third, for agriculture (Mallat 1995). As discussed by Abderrahman (this volume) – although this is not discussed at length in Islamic sources – countries such as Saudi Arabia commonly allocate water to industrial and recreation uses after domestic, animal watering, and agricultural needs have been satisfied.

Obviously, as the population grows and settlement patterns change, that is, as society evolves from rural and agrarian to urban and industrial, reallocation not only is permissible, but also is necessary to preserve equity and the primacy of the right to quench thirst. In theory, the explicit primacy of domestic and animal watering demand over irrigation makes this reallocation easier to support in Muslim countries than in non-Muslim countries. In such a case, the state, which ideally acts as the representative of the people and protects the weak, can and should interfere to determine the utilization priority of water.

Prerequisites for water markets

In parts of the United States, and in Chile, the prerequisites for effective and equitable water markets exist – but do they elsewhere? At a minimum, necessary prerequisites include appropriate legal frameworks, institutions, regulatory mechanisms, economic policies, and infrastructure.

Before water markets are established for reallocation, there must first be clear property rights to water separate from land ownership, and those rights must be tradeable. In addition to Chile and the United States, in recent years, several Australian states, Mexico, and Peru have established property rights to water (Chaudhuri 1996). As noted, clear, tradeable private property rights to water, separate from land, are allowable in Islam. However, this is not necessarily explicit in the secular legislation of some MENA countries.

Also, legislation protecting the environment or third-party water rights from excessive withdrawals must be in place. Protection of the environment and the water rights of others, including plants and animals, is required in Islam – for instance, the Prophet Muhammad (pbuh) states, "there is a reward for serving any animate (living) being,"[1] and "He who digs a well in the desert ... cannot prevent the animals from slaking their thirst at this well."[2] The Majalla code identified *harim* as protected areas – where it is forbidden to dig a well that would endanger the quality or water supply of an existing source. Islam makes one liable for withholding or misusing water, including polluting or degrading clean water – the Prophet Muhammad (pbuh) said, "(Amongst the) three persons whom Allah will not look at on the Day of Resurrection, nor will he purify them, and theirs shall be a severe punishment, [is] a man [who] possessed superfluous water on a way, and he withheld it from travellers."[3]

In addition, institutions that can act as intermediaries between buyers and sellers to enforce fair trading are essential. Water user associations may be able to play a crucial role, even acting as substitutes for formal legal action and serving as pressure groups to enhance the efficiency of the bureaucracy. It may be possible for traditional water-sharing arrangements, distribution networks – such as the Yemeni system of water sharing through "spate" irrigation (small dams built during the flood season in communal co-operation) – and allocation mechanisms to evolve into water rights networks.

The recent academic concept of community-based resource management and participatory development has always existed in Islam. The Prophet's (pbuh) statement that "Muslims have common share in three (things), grass, water, and fire (wood)"[4] implies the right to share in the management of these three common property resources. Also, the Quran describes believers as those "*who (conduct) their affairs by mutual consultation.*"[5] The fulfilment of *shura* (the duty of rulers to consult their peoples) was required even of the Prophet Muhammad.

In concert with regulation to ensure fair trading, some level of government deregulation may be necessary to allow private sector institutions to formally enter the water market, and to allow prices to rise. In fact, although a just ruler regulating the affairs of state to protect the weak is important in Islam, excessive government interference in the market, including the fixing of prices, is frowned upon.

Raising prices for water in urban areas will help simultaneously to reduce the demand of served customers and to provide an economic incentive for intersectoral water markets. There is ample room to raise prices for the served middle and upper classes – water rates in LDCs are typically less than one-sixth the full cost of water provision (Bronsro 1998). The actual full cost of providing water services will vary from country to country, but in Israel, the only country in the MENA where water is charged at full cost in urban areas, the price, including a surcharge for wastewater treatment, is US$1 per cubic metre.

Full-cost pricing is allowable in Islam. In Iran, where the law is based upon *sharia*, irrigation water must be sold on the basis of average cost (comprising both operation and maintenance costs and capital depreciation). This requirement is enshrined in the 1982 Just Distribution of Water Law, the very title of which makes the rationale for full-cost pricing self-evident. For urban areas, a 1990 act allows for full (average) cost recovery, including both capital and depreciation costs. As a result of this bill, in 1996, tariffs were increased by 25–30 per cent where household monthly consumption exceeded 45 cubic metre, and the tariff for commercial and industrial use was set higher than for residential consumption, reversing an earlier policy (Sadr, this volume).

Where does this leave the poor? First, in almost every MENA city, a realistic water price, which would allow for reinvestment into the system to serve the unserved poor, would be less than they currently pay, but higher than that for the serviced urban residents. In the Ivory Coast in 1974, only 30 per cent of the urban population and 10 per cent of the rural population had access to safe water; by 1989, 72 per cent of the urban population and 80 per cent of the rural population (through water points) had access to safe water. The reason for this improvement was that the private water company, the Société de Distribution d'Eau de la Côte d'Ivoire (SODECI), was allowed to increase urban tariffs above the level of long-term marginal costs, especially for industrial customers (Bhattia et al. 1995). Second, tariffs can be structured to supply everyone with a lifeline water volume, as is done in Iran, where about the first 30 LPCD is provided to all households – five thousand litres per household per month, based on an assumed average of six persons per household (Sadr, this volume): this approaches the basic human-need standard of 50 LPCD proposed by Lundqvist and Gleick (1997). Finally, many economists suggest that governments should subsidize income, not water, and this argument finds sympathy in Islam, in which wealth redistribution duties (*zakaat*) are central and compulsory for every Muslim whose wealth qualifies.

Finally, an infrastructure system must be in place to transfer water from buyers to sellers, without inordinate transaction costs. Not all

MENA countries yet have such an infrastructure system in place, but some do, such as Jordan. So also do some non-MENA Muslim countries, such as Pakistan.

Problems and obstacles

The problem is, of course, that the necessary preconditions for the establishment of water markets do not exist in many MENA countries, or in other developing countries. The obstacles to establishing these necessary preconditions transcend water management and include some of the most intractable and difficult development challenges in many countries, such as weak institutions and inequitable access to land and water rights. This has not prevented unregulated intersectoral water markets from springing up in countries such as Bangladesh, Brazil, and India, and in MENA countries such as Jordan and Palestine. In fact, some water trading, if not intersectoral at least intrasectoral, is happening in every MENA city. However, such unplanned and unregulated markets harm third-party interests or the environment. The following discussion illustrates challenges facing equitable and effective development of water markets.

One problem is that lack of government regulation has often resulted in both third-party effects and externalities. Thus, poor farmers in Bihar, India, are selling groundwater to richer farmers or to peri-urban residents for domestic use. State-subsidized cheap electricity has resulted in overpumping, excessive withdrawals in many locations, and falling groundwater tables. This, in turn, has meant that poorer farmers, who cannot sink wells as deep as richer farmers, have lost access to groundwater. Lower groundwater tables will eventually result in streams that are recharged by the groundwater running dry.

Furthermore, despite what Islam may say about the need to ensure equity for the weaker or poorer members of a society, the lack of influence or power of the poor is a common thread in both Muslim and non-Muslim countries. Many authoritarian regimes in Muslim countries certainly do not follow the concept of *shura* by allowing citizens to participate in the planning of projects that affect them. A study that examined the potential for water markets in Jordan recommended strengthening water users' associations, especially in the Highlands and Southern Ghors, to help small farmers by exerting effective pressure on the government and bringing about favourable changes both in policy and services provided (Shatanawi and Al-Jayyousi 1995). On the other hand, the wealthy are liable to have excessive influence on government policies.

Large farmers, in particular, as well as the upper class in urban areas, tend to have very strong lobbies for their interests. Despite its permissibility in Islam, raising water tariffs in this environment is a significant challenge, particularly if the service provided is poor.

Finally, the regulation of externalities related to withdrawals, consumption, changes in water quality, and so forth requires very sophisticated legal and monitoring systems that do not yet exist in most MENA countries. In some cases, laws exist but the government ability to monitor or overcome corruption is weak.

None of these problems are posed by Islam. Rather, they are common to almost all developing countries. In fact, according to Islam, the prerequisites discussed earlier would all be required before water markets for reallocation could be introduced, so as to ensure that they were equitable. Furthermore, some of the necessary legal frameworks, for example, protection of third parties and the environment, were developed in Islamic law before they were appeared in modern Western law.

Food security

Reducing the amount of freshwater available for agriculture understandably raises concerns about national food availability and the socioeconomic impact on poor farmers and farm workers. These are valid concerns about which two points can be made.

First, it is suggested that agriculture should receive water of a different quality, not less of it. An intersectoral transfer policy must be accompanied by increasing urban wastewater treatment, and recycling the same water back to agriculture. For instance, Israel plans to reduce its total freshwater volume allocated to agriculture from 70 per cent in 1996 to 20 per cent by 2030 – in fact, the amount of freshwater left over for agriculture may be even less than 20 per cent if Israel eventually allocates some portions of the freshwater currently under its control to its neighbours so as to achieve a peace agreement (Shuval, as cited in Lundqvist and Gleick 1997, 37). This reduction of freshwater for agriculture will be accompanied by an expansion of wastewater treatment so that 80 per cent of urban wastewater will be treated and recycled back to agriculture – as a result, Israel will have essentially the same amount of water for agriculture as it has at present.

Except in Israel and a few other countries such as Tunisia, only a small percentage of wastewater is treated and reused in the MENA. There are several reasons for this, including water tariffs that do not cover the cost of wastewater treatment, the unsustainability of centralized and highly mechanized plants imported from developed countries, and the notion that wastewater reuse is against Islam. As Abderrahman notes in this volume, according to a *fatwa* of the council of Leading Islamic Scholars in Saudi Arabia, reuse of wastewater is allowable in Islam for virtually any purpose, provided that public health is protected. Saudi Arabia is currently reusing about 20 per cent of its wastewater for irrigating agricultural crops and landscape plants, and in refineries.

Given that people in the Middle East are already largely frugal in their water use, and that freshwater will increasingly be taken away from agriculture, expanding wastewater reuse in agriculture is probably the single most important water demand management policy initiative in the MENA. Also, because safe reuse depends on adequate treatment, it is vital that virtually every drop of wastewater receive at least some treatment. This will require higher water tariffs and a major expansion of wastewater treatment. Treatment plants will consist largely of decentralized low-cost, natural waste treatment systems, for reuse on or near site. The IDRC is currently pilot-testing grey-water treatment using on-site, small-scale, trickling filters for home gardens in the low-density hill settlements surrounding Jerusalem, aquatic wetlands using water lettuce or duckweed in the Jordan Valley and Morocco, and low-mechanical content-activated sludge in Egypt.

The second point to be made regarding food security is that in hard reality, most MENA countries simply do not have sufficient water for national food self-sufficiency in any case. The water scarcity benchmark level of 1,000 cubic metres per person per year (m^3/p/y) includes the amount necessary for food self-sufficiency (Lundqvist and Gleick 1997). As noted, however, the average water availability in the MENA will be 725 m^3/p/y by 2025, and Jordan, Tunisia, and Yemen will all have much less than this. When so little water is available, the first priority must be water for drinking and for domestic purposes, not agriculture. As a result, the concept of food self-sufficiency must give way to national food security (Lundqvist and Gleick 1997), or regional food self-sufficiency, and imports of "virtual water" through the purchase of foods and products produced where it is most efficient. In addition to Israel, water-scarce countries such as Botswana have already accepted this fact, and the latter does not have a policy of food self-sufficiency but tries to ensure food security by annual negotiations with suppliers of cereals. Shuval (cited in Lundqvist and Gleick 1997) suggests that a small amount of fresh water, 25 m^3/p/y, should be reserved for domestic production of fresh vegetables, which have high economic and nutritional value. Some of this production may come from the growing practice of urban agriculture – intensive vegetable production may use as little as 20 per cent of the water and 17 per cent of the land, required for rural, tractor-cultivated crops (UNDP, cited in Lundqvist and Gleick 1997, 25). Such urban garden vegetables will usually be cheaper for the poor than imported ones. Where feasible, most other crops in arid countries will have to be grown, increasingly and eventually exclusively, with treated wastewater.

Integrated water resources management

Intersectoral transfers are not an end in themselves, but rather a necessary tool to balance the benefits flowing from water across a society. When not enough water is available for all potential uses, hard choices must be made about which sectors, activities, and regions should receive water, and how much each should receive. MENA governments have to view water as a precious national resource and put in place a process to allocate water using an integrated water management approach that acknowledges the interdependence of all water issues. Among other components, this process requires multi-stake-holder decision-making, macrolevel modelling, and co-operation between government departments.

The allocation process will have to begin at the local level, so that all affected stake-holders have a voice, and allocation priorities are not set only by those who are most powerful. Some will lose their water, and this will lead to inequities, at least in the short term. But equity implications across regions and sectors can only be analysed at a national level, if demands from various sectors and regions are fed into a national allocation process. The process itself will probably be iterative, but once allocations for a given basin are determined, water management should be decentralized, and decisions made at the lowest appropriate level. Furthermore, these allocations will have to be reviewed periodically, as conditions in a country change. Countries such as Israel are moving toward a system where withdrawal permits are granted only on a short-term basis, and the renewal of permits will be subject to the government's periodic assessment of the best overall use of water in the country.

A good example of the need for macro-level modelling is provided by a recent World Bank study in Algeria, which found that a proposed irrigation project was in direct competition for the same water with another proposed project for urban water supply (Rogers 1993). In such a case, the relative marginal benefits to the national economy (including the effect on the poor) of additional investment in irrigation must be carefully compared to the benefits of investment in the urban sector. This, in turn, requires estimates of the opportunity cost of different water qualities, the short- and long-term effects of dislocating small-scale farmers and farm workers, and the alternative employment opportunities. Other resource sectors, such as energy, have well-developed methodologies to relate sectoral and macro plans. A few macro-level plans have been developed for water allocation, but they have been used only sporadically (Rogers 1993).

Many, if not most, nations in the MENA face the inevitable decision to move from a policy of food self-sufficiency to a policy of food security. This requires both external and internal integration. In order to purchase food grown elsewhere in the region, states must be able to earn enough foreign currency from industrial exports and tourism, and must have stable trading relationships. In addition to external co-operation, this requires integrated internal policies based upon co-operation among government departments, including agriculture, trade, tourism, and industry.

Governments need to set a vision for national water allocation, and to regulate markets so that transfers will be slow, constant, and thoughtful. Using the values quoted earlier, if we assume that 100 units of renewable water are available to a country as a whole, transferring 8 units from agriculture requires only a 10 per cent increase in sectoral efficiency, but nearly doubles the amount available for domestic purposes: and this leaves out of account the possibility that the same volume might be returned to irrigation as treated wastewater. In fact, demand management in rural areas is far more likely if users have an economic incentive to voluntarily trade their water use rights. Also, it has been proven that it is not only possible to maintain agricultural production but even to increase it while reducing water use, especially when beginning with the low-efficiency irrigation practices common in most MENA countries. Thus in Africa, increases in agricultural production have been achieved in Kenya (Machakos) and Niger (Keita) while reducing the use of water or reversing land degradation (Templeton and Scherr 1997).

Conclusion

The scarcity of water in the MENA region is becoming critical, and the very high rates of urbanization are pressuring governments to transfer water out of rural areas, where most of it is used, to urban areas, where the majority of the region's absolute poor already live. Alongside the inequities in access to water in rural areas is the growing inequity in urban areas, where the unserved poor pay very high prices in informal intra-urban water markets. Also, as the amount of water available to cities decreases per capita, the situation of the urban poor will further deteriorate.

Along with demand management measures to optimize the available water in urban areas, intersectoral water markets have been suggested as a means to transfer freshwater from farmers in rural areas (who willingly sell it) to urban areas. This must be accompanied by expanded wastewater treatment and reuse in agriculture, particularly in peri-urban agriculture. Regulated water markets have been very successful in developed countries such as the United States and Chile, and intersectoral transfers through water markets are inevitable in the MENA as well. Already, the growing scarcity of water and its high black-market price have resulted in unregulated water markets in such MENA countries as Jordan, Lebanon, and Palestine. Unregulated markets without necessary legal, institutional, and economic measures in place can lead to unsustainable practices such as in India, where groundwater tables have dropped alarmingly as a result of farmers selling their water to other farmers or to cities.

For most categories of water, trading is permissible according to Islam. Indeed, given the very clear guidelines on the priority of water rights in Islam, some reallocation of freshwater from rural to urban sectors is not only allowable, it is desirable. Accordingly, MENA governments must undertake studies, make necessary legal, institutional, and economic reforms, and establish a process incorporating integrated water management and multi-stake-holder input to thoughtfully allocate water to meet societal goals. Governments must also consider providing employment opportunities to farmers and farm workers. Without such leadership measures, growing unregulated markets will result in still greater inequities because, increasingly, water will flow primarily to the rich and powerful, with little left for the poor and marginalized.

Notes

1. Al-Bukhari 8.38.
2. Al-Bukhari 5550, in *Hadith Encyclopedia.*
3. Al-Bukhari 3.547.
4. Abu-Dawood 3470.
5. 26:38.

REFERENCES

Bhattia, R., Cesti, R., and Winpenny, J. (1995), *Water Conservation and Reallocation: Best Practice Cases in Improving Economic Efficiency and Environmental Quality*, Joint Study, World Bank–Overseas Development Institute, Washington, D.C.

Bhatia, R. and Falkenmark, M. (1993), *Water Resources Policies and Urban Poor: Innovative Approaches and Policy Imperatives, Water and Sanitation Currents*, UNDP–World Bank Water and Sanitation Programme, Washington, D.C.

Bino, M. J. and Al-Beiruti, Shihab N. (1998), "Inter-Islamic Network on Water Resources Development and Management (INWRDAM)," *INWRDAM Newsletter* 28 (October).

Bronsro, A. (1998), "Pricing Urban Water As a Scarce Resource: Lessons from Cities around the World," in *Proceedings of the CWRA Annual Conference, Victoria, B.C., Canada*, Canadian Water Resources Association, Cambridge, Ont.

Chaudhuri, S. (1996), "To Sell a Resource," *Down to Earth*, 15 February 1996, pp. 35–37.

Lundqvist, Jan and Gleick, Peter (1997), *Comprehensive Assessment of the Freshwater Resources of the World – Sustaining Our Waters into the 21st Century*, Stockholm Environment Institute, Stockholm.

Mallat, Chibli (1995), "The quest for water use principles," in M. A. Allan and Chibli Mallat (eds.), *Water in the Middle East*, I. B. Tauris, New York.

Postel, S. (1995), "Waters of Strife," *Water* 27 (November–December), pp. 19–24.

Rogers, Peter (1993), "Integrated Urban Water Resources Management," *Natural Resources Forum* 10 (February), pp. 33–42.

Shatanawi, M. R. and Al-Jayyousi, O. (1995), "Evaluating Market-Oriented Water Policies in Jordan: A Comparative Study," *Water International* 20 (2), pp. 88–97.

Templeton, S. R. and Scherr, S. J. (1997), *Population Pressure and the Microeconomy of Land Management in Hills and Mountains of Developing Countries*, Discussion Paper 26, Environment and Production Technology Division, International Food Policy Research Institute, Washington, D.C.

12

Management of shared waters: A comparison of international and Islamic law

Iyad Hussein and Odeh Al-Jayyousi

International water resources include surface water such as rivers and lakes and their tributaries, and groundwater such as aquifers and ground basins, that lie within the jurisdiction of two or more states. Management of such shared water resources must take many factors into account, including current laws, existing legal and institutional frameworks, present and future water resources and uses, climatic conditions and water availability in the basin or region concerned, water cost from different sources, and users' ability to pay. This chapter argues that the conceptual legal framework which already exists in theory might serve in the management of shared water resources. However, on international water issues, no one law is universally applicable.

Moore (1992) argues that in the realm of international water law, there is no universally accepted definition of equity in the division of waters between users. Because the characteristics of each international water resource are specific in hydrologic, institutional, and legal aspects, rules and regulations of universal applicability are not realistic unless they are kept broad and flexible. Conflicts have arisen over the years between riparian states with a shared water resource because of different approaches to the issue of sovereignty. The conflict is always between upstream and downstream states.

International water law and practice

National water policy is more likely to be influenced by a country's upstream or downstream position within a basin than by international law. The only constraint is the fear of setting unfavourable precedents in further dealings with neighbouring countries and the disapproval of the international community.

Traditionally, there are five theories governing the use of international rivers (Utton and Teclaff 1978):

- Absolute territorial sovereignty (the Harmon Doctrine), which ascribes to upper riparian states absolute sovereignty over rivers flowing through their territory;
- Absolute territorial integrity, which guarantees to lower riparian states the use of rivers in an unaltered condition;
- Limited territorial sovereignty or equitable utilization theory, which permits use of rivers so far as no harm is done to other riparian states;
- Limited territorial integrity, which recognizes the existence of a community of interests among riparian states that gives rise to a series of reciprocal rights and obligations; and
- Drainage basin development or the community of interests theory, which stresses common development of rivers by all riparian states.

The last theory has become the most widely advocated by the international legal community (Utton and Teclaff 1978). The community of interests theory recognizes that both upstream and downstream states have a legitimate interest in water resources and tries to balance their use to the mutual benefit of all parties concerned (Wilson 1996). In 1966, the International Law Association (ILA) formulated the Helsinki Rules on the Uses of Waters of International Rivers, which embodied this concept and adopted the notion of equitable utilization.

The same concept was adopted by the International Law Commission (ILC) of the United Nations (UN) in 1991 in the Draft Articles on the Law of the Non-Navigational Uses of International Watercourses. These draft rules were reviewed by UN member states' governments and experts in the field and were reassessed in light of these comments in the 1993 and 1994 meetings of the ILC. The commission finally adopted a text containing thirty-three articles in the summer of 1994 and submitted them to the General Assembly, which approved them in May 1997 by Resolution 51/229.

The main concepts and principles included in the ILC articles (ILC 1997) may be summarized as follows:

The articles aim to achieve a balance between the "equitable and reasonable" utilization of an international river by any individual riparian state (article 5) on the one hand, and on the other hand the desirability of avoiding "significant harm" to other riparian states that are already using the river (article 7), or might want to use it in the future. The articles stress the riparian states' obligation to protect international rivers and associated ecosystems (articles 5, 8, 20, and 21). They oblige riparian states to co-operate in the optimal utilization and protection of the rivers that they share (article 8) and recognize that agreements between riparian states may cover the entire river basin, or only part of it (article 3). In the latter case, however, the agreement should not "adversely affect," to a "significant extent," other riparian states' use of the waters in the basin. The first paragraph of article 7 reads: "Watercourse states shall, in utilizing an international watercourse in their territories, take all appropriate measures to prevent the causing of significant harm to other watercourse states." Article 10, on the relationship between different kinds of uses, reads: "In the absence of agreement or custom to the contrary, no use of an international watercourse enjoys inherent priority over other uses."

The type of issues, constraints, and opportunities that prevail in the management of shared water resources appear clearly, in the case of rivers, in the history and background of the Nile Waters Agreement (Flint 1995), and in the case of underground resources, in the current issues between Jordan and Saudi Arabia involving the Rum Aquifer (Naff and Matson 1984).

The catchment area of the Nile is shared by eight states between its source and the Mediterranean Sea: Rwanda, Uganda, Tanzania, Kenya, Zaire, Ethiopia, Sudan, and Egypt. The leading riparian state in terms of political and physical influence over the Nile is Egypt. The upstream states of both the Blue and White Niles are in a weak position because of political and economical instability. During the early years of Egyptian independence, downstream dominance was maintained in negotiations with the British over Sudan's water usage. In 1929, an agreement was reached between Egypt and Sudan that allocated the water of the Nile between these two parties. The Nile Waters Agreement was reassessed and finalized in 1959.

In July 1993, a general agreement grounded in international law was reached by the downstream countries with the new Ethiopian government which may mark the beginning of a new era of co-operation. This agreement included a clause that the upstream countries agreed not to act in a way that might harm the downstream states, and to consult and co-operate on future water projects that would be mutually beneficial.

It can be concluded that the riparian states of the Nile River will be looking to new developments in international water law including the ILC's study of watercourses for a way to achieve co-operation and co-ordination in the future.

The area of shared groundwater resources in this case study is the Rum Aquifer, which stretches four hundred kilometres from near Tebuk in Saudi Arabia northward across Jordan to the northern tip of the Dead Sea. Exploitation of fossil water of the Rum aquifer amounts to "drawing on capital" and it would be prudent to manage the rate and duration of this exploitation carefully and to work toward alternative, long-term, and sustainable resource development. At present, the issue of sustainable development is complicated by the substantial exploitation of this water resource in the Tebuk region, and by Jordanian plans involving use of the aquifer. In one evaluation of alternative sources to meet long-term national demand, an annual yield from the Rum aquifer of 50–70 million cubic metres per year (MCM/y) was identified as an economic resource to supply Amman. Other studies approach the required yield of the Rum aquifer from a resource balance point of view and indicate that the one-hundred-year safe yield would be only 110 MCM/y (Thames Water 1988).

At present, Saudi Arabia's usage of the aquifer is much greater than Jordan's. The two countries have held discussions but no agreement has been reached. The lack of a joint legal and institutional mechanism between Jordan and Saudi Arabia necessitates the adoption of a joint water agreement based on international or Islamic law principles.

International water law and Islamic water law principles

Some Islamic maxims that have specific implications for water planning and management are those involving equitable and reasonable utilization, ownership, significant harm, the duty of consultation, and preserving the environment and ecosystem. These are discussed in the following sections in relation to the main principles of international water law, with emphasis on the ILC Articles.

Equitable and reasonable utilization

In Islam, beneficial use of water is best viewed through the broad provisions against misuse of rights. Use of rights is governed by moral and legal regulations. The former require good conduct and consideration for others as well as conformity to accepted norms.

Waste of any kind is forbidden by Islamic law, above all in respect of water. Muslim jurists stipulate that every individual has the right to benefit from something that is *mubah*, that is, free of all restrictions of ownership and conditions that in any way undermine its availability for all humankind. Thus, people make use of rivers and lakes that are not owned just as they make use of air and light. Although water may be used for a variety of purposes, the user is not free to dispose of it or to benefit from it in a manner detrimental to others.

Large expanses of water that present no problem in allocation of water are shared equally. Small streams or lakes are allocated in the first instance to those dwelling in the vicinity of the water source. If, however, the water is not present in sufficient quantity to satisfy everyone's needs, water rights are allocated in the following manner:

- Where the water of the stream or source requires no artificial means to be extracted, those nearer to the source take first and then those who are further away, and so on: those occupying higher ground have priority over those occupying lower ground.
- Where effort is required to make the water flow, allocation is determined on the basis of several factors, including the expense and labor contributed by each state, its population, and its domestic, agricultural, and industrial demands.

In both cases, the Islamic law provisions regarding the provision of surplus water to others in need strictly applies.

More generally, Islamic law recognizes the following priorities for water use.

- *Haq al shafa* or *shirb* – the right to quench thirst;
- Domestic use, including watering animals;
- Irrigation of agricultural lands; and
- Commercial and industrial purposes.

In international law, there is no accepted definition of equity. However, the Helsinki Rules on the Uses of the Water of International Rivers identify several factors thought to have a bearing upon equity. Chapter 2 of the ILC Rules deals with equitable utilization of the waters of an international drainage basin in article 5: "Watercourse states shall in their respective territories utilize an international watercourse in an equitable and reasonable manner. In particular, an international watercourse shall be used and developed by watercourse states with a view to attaining optimal and sustainable utilization thereof and benefits therefrom, taking into account the interests of the watercourse states concerned, consistent with adequate protection of the watercourse."

The factors that should be taken into account when determining a reasonable share of basin waters for each basin state are the geography of the basin, its hydrology and climate, past utilization of its waters, the economic and social needs of each basin state, the population dependent on the waters of the basin in each state, the comparative cost of alternative means of satisfying the economic and social needs of each state, the availability of other resources, avoidance of unnecessary waste in the utilization of the waters of the basin, and the degree to which the needs of a basin state may be satisfied, without causing substantial injury to any other basin state.

Ownership of water

The ownership of water is addressed in the chapters in this volume by Kadouri, Djebbar, and Nehdi and by Caponera. As discussed there, any control of water that does not involve possession in the strictest sense – that is, storage in a cistern or, pool, or by any other means that contains the water within well-defined bounds – is not to be regarded as ownership. Thus, although wells and artificial springs may be claimed as private property, the water in them is never regarded as private property until it comes into "possession" in this sense (*hiyazah*).

It is, therefore, generally understood in Islamic law that, although the right of the public to benefit from *mubah* water is well established, the water nevertheless remains under the supervision and direct protection of the law. In this sense, any member of the public may seek a judicial decision to establish a water right or to protect it. Such claims may be instituted against a person claiming private ownership of water or a member of the public preventing him from use of water.

Likewise, ownership of water under international water law consistently accepts that states have the sovereign right to explore and exploit their own natural resources. However, there is a balancing obligation upon states to be aware of the trans-boundary effects of their activities, and to be liable for pollution for which they are responsible.

Significant harm and compensation

In Islam, a well-known *hadith* of the Prophet Mohammed (pbuh) states: "Don't commit any harm or injury to yourself, and do not cause harm or injury to others."[1] In line with this, Islamic law gives priority to public interest, and to observance of the following principles.

- Harmful practices must be eradicated;
- Harmful practices may be tolerated only where they prevent the use of other practices that are considered to be more harmful; and
- It is advisable to prevent harm rather than to provide benefits.

Islamic law is implemented either directly by supervised application or by judicial remedy. Therefore, waters that fall under the general category of public ownership are directly supervised by the government and all provisions relating to them are implemented by government officials. Punishment for contravening these provisions is through either a jail term or a fine – but more often through a fine.

Under international water law, article 7 of the Final Articles of the ILC provides that watercourse states must take all appropriate measures to ensure that their activities do not cause significant harm to other such states. Further, where significant harm is caused, the state that causes the harm is obligated to consult with the state suffering the harm to determine whether the use that is responsible for the harm is reasonable and equitable, to make ad hoc adjustments to the use to eliminate or mitigate the harm, and to make compensation, where appropriate.

Consultation

In Islam, *shura*, or consulting the public, is one of the bases for decision-making by the government and its officials. Muslims believe that God ordered the Prophet Muhammad (pbuh) to consult people in public matters before taking a decision. In international water law, states also have a duty to consult their neighbours if they intend to exploit a trans-boundary water resource and there is potential that the activity will have a trans-boundary effect.

Preserving the environment and ecosystem

The importance in Islam of preservation of the environment is established by Amery in this volume. Another example of the Islamic stress on the environment is the *hadith*, "A Muslim does not plant a sapling but a man or an animal or a bird eats of it, it is a charity for him till the Day of Resurrection."[2] In the same way, the ILC Articles provide that watercourse states shall individually or jointly protect and preserve the ecosystem of an international watercourse (article 20), and prevent, reduce, and control pollution of the watercourse (article 21).

Conclusions

From the foregoing comparison of international water law and Islamic water principles, it can be concluded that a number of common bases exist and a mutual approach can be established. Reasonable shares, equity, public interest, consulting, and preserving the public interest and the ecosystem are the main elements that overlap. However, there is a lack of literature on Islamic perspectives related to shared waters, and further work is needed to develop an Islamic water management policy that covers shared waters.

It is recommended that a workshop should be organized between Muslim scholars and water experts in the Muslim world to develop a consensus with regard to the position of Islamic law on shared waters. Afterwards, a consultative council of selected experts with jurists, scholars, and water experts from the Muslim world should be established to set up Islamic water policies and formalize an Islamic water law. Once a basis for the Islamic law on shared waters is established, a pilot project can be implemented, covering different cases in the Muslim countries, to translate the theory into actions in the real world.

Note

1. Al Baghdadi 32, in Al Baghdadi 1982.
2. An-Nawawi 135, in An-Nawawi 1983.

REFERENCES

Al Baghdadi, Abu Abd Al Rahman Mohammed bin Hasan (1982), *Jamma Al Aloum Wal Hikam* [Collection of the sciences and wisdom] (5th ed.), Dar Al Manhal, Cairo.

An-Nawawi, Yahia Ibn Sharaf (1983), *Riyadh-Us-Saleheen* [The garden of the righteous], trans. S. M. Abbasi, vol. 1, Dar Ahya us Sunnah, Al Nabawiya, Karachi.

Flint, C. G. (1995), "Recent Development of the International Law Commission Regarding International Watercourses and Their Implications for the Nile River," *Water International* 20, pp. 197–204.

ILC (International Law Commission) (1997), *Convention on the Law of the Non-navigational Uses of International Watercourses*, United Nations, General Assembly Resolution 51/229, United Nations, New York.

Moore, J. (1992), *Water Sharing Regimes in Israel and the Occupied Territories – A Technical Analysis*, Project Report 609, Operational Research and Analysis Establishment, Department of National Defense, Ottawa.

Naff, T. and Matson, R. (1984), *Water in the Middle East: Conflict or Coordination?* Westview Press, Boulder, Colo.

Thames Water (1988), *Water Quality in Greater Amman Study*, Ministry of Planning, Amman.

Utton, A. E. and Teclaff, L. (1978), *Water in a Developing World: The Management of a Critical Resource*, Western Special Studies in Natural Resources and Energy Management, United Nations Development Programme, New York.

Wilson, P. (1996), *The International Law of Shared Water Resources*. Training Manual on Environmental Law, United Nations Environment Program, Nairobi.

Glossary of Arabic and Islamic terms

The definitions given in this glossary are presented in the context of this volume. Many of the terms may have wider meanings. Only those terms cited in this volume are included.

akhlasu. Genuine in religious beliefs.
al-hawa. Personal temptations.
Allah. The Arabic word for the One True God, in which **Muslims** believe, who created and sustains all life. The word *Allah* is unique in that it has no plural nor any gender connotation.
anfal. Property of the Imam, the just and legitimate ruler.
baghi. Rebel, oppressor, or transgressor.
baitulmal. Public treasury.
bid'ah sayyi'ah. Inquiry prohibited in **Islam**.
djahilyya. Period of ignorance that **Muslims** believe preceded the arrival of **Islam** in Arabia.
dhimmis. Non-Muslims who live in an **Islamic** state.
fahesha. Despoiling; committing a bad, shameful deed, including despoiling of natural resources (plural *fawhish*).
faqih. One who has a deep understanding of **Islam**, its laws, and jurisprudence; a jurist (plural *fuqaha*).
fassad. Corruption, chaos, mischief, or spoiling of anything including water resources.
fatahna. Literally "opened up" (or "poured") blessings such as ***rizq***, which includes natural resources such as water.

fatwa. Legal ruling on an issue of religious importance (plural *fataawaa*).

fenjan. Local unit of water in some parts of Iran; Arabic for "cup."

fiqh. Literally "comprehension" or "knowing." The branch of learning concerned with the injunctions of the ***sharia*** relating to human actions, derived from the detailed evidence pertaining to them. See ***faqih***.

ghusl. Purifying bath that **Muslims** must take after conjugal relations or prior to offering prayer.

hadith. A narration describing what the Prophet (pbuh) said, did, or tacitly approved.

hadith qudsi. A special category of ***hadith***, which contains the word of **Allah**, related by the Prophet (pbuh). *Hadith qudsi* differs from the **Quran** in that the latter comprises the exact words of God.

haez. Literally "one who has or owns" anything, including water. See ***hiazat***.

Hanbali. One of the main schools of thought in **Islam**, founded by Ahmad Ibn Hanbal (d. 855 A.C.), a famous scholar of ***fiqh***.

Hanifi. One of the main schools of thought in **Islam**, founded by Abu Hanifah (d. 767 A.C.), a famous scholar of ***fiqh***.

haq al shafa (shirb). Literally "the law of thirst" or "the right to quench thirst"; *shafa* also means "to heal," or "the restoration of health."

haraam. Forbidden in **Islam**.

haram. Sacred or inviolable. A protected area in which bad behaviour is forbidden and other good behaviours are essential. The area around the Ka'ba in Mecca, and the area around the Prophet's mosque in Medina, are *haram*.

harim. Protected (from ***haram***): land surrounding canals, wells, and other water sources on which digging a new well is forbidden so as to protect the quality and quantity of the water source.

hiazat. Spending time and effort to take something into possession. In the context of water, it refers to the effort of supplying, storing, treating, and distributing water (singular *hiyazah*). See ***haez***.

hisba. Office of accounting or public inspection. See ***muhtasib***.

huda. Guidance or direction given by God to His creation.

hudud. Consensus of the jurists since the death of the Prophet (pbuh) on any issue of ***fiqh***.

ijma. Unanimous agreement of Muslim jurists.

ijtihad. Literally "striving and self-exertion: independent reasoning; analytical thought": *ijtihad* is the interpretation of the source materials, inference of rules from them, or giving a legal verdict or decision on any issue on which there is no specific guidance in the **Quran** and the ***sunnah***.

Islam. An Arabic word derived from the root words *silm* and *salaam,* which mean "peace." The meanings of "Islam" include: peace, greeting, salutation, obedience, loyalty, allegiance, and submission to the will of God.

istihsan. Literally "preference": juristic preference where no ruling exists.

istishab. The presumption that ***fiqh*** laws applicable to certain conditions remain valid until proven otherwise.

istislah. Improving or rehabilitating something: literally "seeking the welfare." The principle of ***istihsan*** applied in a more restricted form, which means seeking that which is more suitable to human welfare than some existing condition – for instance, the improvement of land from being idle and waste to being productive.

itaqu. The faithful who fear God.

jizya. Literally "tribute": the tax paid by ***dhimmis*** in an Islamic state in exchange for the protection provided to them and for their exemption from military service and payment of ***zakaat***.

joraeh. Unit of volume used for water in Iran.

khalifa. Viceregent, successor, steward, or trustee (plural *khulafa*). Humans are referred to as the *khulafa* or stewards of God on earth. The word *khalifa* was used after the death of the noble Prophet Muhammad to refer to his successor, Abu Bakr, as the head of the Muslim community. Later, it came to be accepted as the office of the head of the Islamic state. As trustees, the role of humans, especially leaders (*khulafa*) on earth is to ensure that all resources, including water, are used in a reasonable, equitable, and sustainable manner.

kharaj. A tax levied on conquered land by Khalif Umar. This land was not given as booty to the victorious army, but was left to the conquered owners in return for the payment of a tax on the land.

khazen. Finance manager.

khutba. A sermon given by an imam in a mosque before the Friday congregational prayer (*salatul-Jum'ah*).

ma'. Water.

madrasa. Religious school (plural *madaris*).

ma-li. "My money": *maal* signifies wealth or money.

Maliki. One of the main schools of thought in **Islam**, founded by Malik Ibn Anas al-Asbahi (d. 795 A.C.), a famous scholar of ***fiqh***.

maslaha. The public interest. It is generally held that the principal objective of the ***sharia*** is to realize the genuine *maslaha* or benefit of the people.

maulvis. An honorific title of local **Muslim** leaders or Imams in India and Pakistan. The Urdu version of Arabic *mawla* (pl. *mawali* – Persian *mulla*): master, patron, or client. A designation of **Allah** as "*the Protector*" (8:40; 47:11). Now used as a title for religious or political authorities.

mawat, mewat. Dead, idle, or fallow lands.

Mejelle. Literally "magazine": a publication of the Ottoman Civil Code in the 1870s.

miri. Collective ownership: public lands or state-owned land.

mubah. Free of ownership or conditions which undermine the availability for all humankind of the resource so described. For example, air, light, and water in its natural state, such as in the form of precipitations, glaciers, or large lakes, are all *mubah*.

muhtasib. The officer in charge of the ***hisba***, whose duty, among other things, is to ensure the proper conduct of people in their public activities.

mujtahid. Religiously learned.

mulk. Private property with full right of disposal.

munkar. An act that is despised by **Allah**.

mushaa. A form of collective ownership of land.

Muslim. In a general sense, anyone who submits to God, including all the prophets in whom Muslims believe. In a more specific sense, one who submits to God by following the religion of **Islam**.

mustalah al hadith. The science of **hadith** criticism.

nath'ubet. Became scarce.

qiyas. Analogy or relevance.

Quran. The Holy Book, which **Muslims** believe contains the exact revelations made by **Allah** to the Prophet Muhammad (pbuh) through the Angel Gabriel.

rizq. Provisions that God destines for a person, in such forms as additional income, food, clothing, or a natural resource such as water. See ***fatahna***.

sahaba. The companions of the Prophet (pbuh).

salaah, salaat. The prayer that must be offered at least five times a day by every adult **Muslim**.

salam. Peace and harmony of people between themselves or with nature. It is a key word in an Islamic greeting. See **Islam**.

Shafi'i. One of the main schools of thought in **Islam**, founded by Abbas ibn Uthman ibn Shafi' (d. 820 A.C.), a famous scholar of ***fiqh***.

shahed. Witness.

shaqa. Misery.

sharia. The Islamic way of life, Islamic law.

shura. Consultation among decision-makers and between decision-makers and the general public.

sunan. Laws. The plural of ***sunnah***.

sunnah. The way of the Prophet (pbuh): *sunnah* comprises what the Prophet (pbuh) said, did, and encouraged both explicitly and implicitly.

taqua. The faithful who fear God (from the verb *itaqu*).

tatghou. Excess, oppression. *La tatghou*: to commit no excess.
usher. Literally "tenth": a tax representing a percentage of the harvest payable by a **Muslim**. It represents a religious obligation, like ***zakaat***, intended for the benefit of the poor and the needy.
wajeb. A duty that is essential, but not quite compulsory, for **Muslims**.
waqf. An endowment of money or property: the return or yield is typically dedicated toward a certain end, for example, to the maintenance of the poor, a family, a village, or a mosque.
wudu. Ablution: the ritual cleaning of the body before beginning an act of worship.
zakaat. Literally "the purification (of wealth)": payment made by every Muslim who can afford it that is given to the poor and needy. One of the Five Pillars of Islam.

Volume editors

Naser I. Faruqui is a Senior Program Specialist with the International Development Research Centre, Ottawa, Canada, focusing on water management in the Middle East. He was selected by the International Water Resources Association as one of fourteen water leaders of the next generation, worldwide.

Asit K. Biswas is President of the Third World Centre for Water Management in Mexico City. He is a member of the World Commission on Water and Past President of the International Water Resources Association.

Murad J. Bino is the Executive Director of the Inter-Islamic Network on Water Resources Development and Management, Amman. He was previously the Director of the Environment Research Centre of the Royal Scientific Society, Amman.

Workshop participants

Abdul Karim Al-Fusail: National Water Resources Authority, P.O. Box 8944, Sana'a, Yemen. Telefax: 231-530.

Asit K. Biswas: International Water Resources Association (CIIEMAD-IPN), Viveros de Tlalnepantla No. 11, Viveros de la Loma, Tlalnepantla, Edo. de México, 54080 Mexico. Email: *akbiswas@internet.com.mx* or *akb@pumas.iingen.unam.mx*. Tel./Fax: 52-5-754-8604. Switchboard: 52-5-752-0818/586-0838/586-9370.

Cecilia Tortajada: International Water Resources Association (CIIEMAD-IPN), Viveros de Tlalnepantla No. 11, Viveros de la Loma, Tlalnepantla, Edo. de México, 54080 Mexico. Email: *cquiroz@vmredipn.ipn.mx; akb@pumas.iingen.unam.mx*. Tel./Fax: 52-5-754-8604. Switchboard: 52-5-752-0818/586-0838/586-9370.

Dante A. Caponera: Former Chief, UN/FAO Legislation Branch; Chairman, Executive Council, International Association for Water Law; Consultant on Natural Resources, Water and Environmental Law, Via Montevideo 5, 00198 Rome, Italy. Email: *caponera@libero.it*. Tel./Fax: 39-6 8548932.

Dina Craissati: Senior Program Officer, Middle East and North Africa Regional Office, International Development Research Centre, 3 Amman Square (5th floor), Dokki, Cairo, P.O. Box 14 Orman, Giza, Egypt. Email: *dcraissati@idrc.org.eg*. Tel.: 20-2-336-7051/52/53. Fax: 20-2-336-7056.

Ellysar Baroudy: Coordinator, Water Demand Management Research Network, Middle East and North Africa Regional Office, International Development Research Centre, 3 Amman Square (5th floor), Dokki, Cairo, P.O. Box 14 Orman, Giza, Egypt. Email: *ebaroudy@idrc.org.eg*. Tel.: 20-2-336-7051/52/53. Fax: 20-2-336-7056.

Hussein A. Amery: Division of Liberal Arts and International Studies, Colorado School of Mines, Golden, CO 80401-1887 USA. Email: *hamery@mines.edu*. Tel.: (303) 273-3944. Fax: (303)-273-3751.

Iyad Hussein: Business Development Manager, Jordanian Consulting Engineer Co., P.O. Box 926963, Amman 11183, Jordan, and Applied Science University, College of Engineering, Amman 11110, Jordan. Email: *riverside@hotmail.com*. Tel.: (o) 962-6-560-6150/568-7369, (r) 962-6-515-6099. Fax: 962-6-568-2150.

Karim Allaoui: Office of the Vice President (Operations), Islamic Development Bank, P.O. Box 5925, Jeddah 21432, Saudi Arabia. Email: kallaoui@*isdb.org.sa*. Tel.: 966-2-636-1400 ext. 6729. Fax: 966-2-636-6871.

Kazem Sadr: School of Economics and Political Science, Shahid Beheshti University, Tehran 19834, Iran. Tel.: (o) 98-21-240-3020, (r) 98-21-808-3844. Fax: 98-21-880-8382.

Murad Jabay Bino: Executive Director, Inter-Islamic Network on Water Resources Development and Management (INWRDAM), P.O. Box 1460, Jubayha, Amman 11941, Jordan. Email: *inwrdam@amra.nic.gov.jo*. Tel.: 962-6-533-2993. Fax: 962-6-533-2969.

Nader Al Khateeb: Water and Environmental Development Organization (WEDO), P.O. Box 844, Bethlehem, Palestine. Email: *wedo@p-ol.com*. Tel.: 972-2-747948. Fax: 972-2-745968.

Naser Irshad Faruqui: Senior Program Officer, Water and Wastewater Projects, Program Branch, International Development Research Centre, P.O. Box 8500, Ottawa, ON, Canada, K1G 3H9. Email: *nfaruqui@idrc.ca*. Tel.: 613-236-6163 ext. 2321. Fax: 613-567-7749.

Odeh Al-Jayyousi: Applied Science University, College of Engineering, Civil Engineering Department, Amman 11931, Jordan. Email: *jayousi@go.com.jo*. Tel.:(o) 962-6-5237181, (r) 962-6-5851809. Fax: 962-6-5232899.

Sadok Atallah: Director, Environmental Health Programme, World Health Organization (WHO), P.O. Box 1517, Alexandria 21511, Egypt. Email: *ceha@who-ceha.org.jo*. Tel.: 203-482-0223. Fax: 203-483-8916. (Dr. Atallah is currently on leave from WHO. He can be contacted in Tunis at: Email: *baby.world@planet.tn*. Tel.: 216-1-887263. Fax: 216-1-238182.)

Saeeda Khan: Former Workshop Coordinator, IDRC, 51 Westfield Cres., Nepean, ON, Canada, K2G 0T6. Email: *thekhans@home.com*. Tel.: 613-820-0682.

Shihab Najib Al-Beiruti: Head of Services and Programs Section, Inter-Islamic Network on Water Resources Development and Management (INWRDAM), P.O. Box 1460, Jubayha, Amman 11941, Jordan. Email: *inwrdam@amra.nic.gov.jo*. Tel.: 962-6-533-2993. Fax: 962-6-533-2969.

S. M. Saeed Shah: Head of Hydrology Division, Centre of Excellence in Water Resources Engineering, University of Engineering and Technology, Lahore 54890, Pakistan. Email: *centre@cewre.lhr.sdnpk.org*. Tel.: 92-42-682-2024/1100.

Walid A. Abderrahman: Manager, Water Section Center for Environment and Water Research Institute, King Fahd University of Petroleum and Minerals (KFUPM), P.O. Box 493, Dhahran 31261, Saudi Arabia. Email: *awalid@kfupm.edu.sa*. Tel.: (o) 03-860-2895, (r) 03-860-6962. Fax: 03-860-4518.

Yassine Djebbar: Project Engineer, South Areas Division, Sewage and Drainage Department, Greater Vancouver Regional District, 4330 Kingsway, Burnaby, BC, Canada, V5H 4G8. Email: *yassine.djebbar@gvrd.bc.ca*. Tel.: 604-451-6144. Fax: 604-436-6960.

Index

CPSIA information can be obtained
at www.ICGtesting.com
Printed in the USA
LVHW102347240723
753342LV00004B/62